The Essential Guide to the Business of U.S. Mobile Wireless Communications

ISBN 0-13-042055-7

Essential Guide Series

THE ESSENTIAL GUIDE TO DATA WAREHOUSING
Agosta

THE ESSENTIAL GUIDE TO WEB STRATEGY FOR ENTREPRENEURS
Bergman

THE ESSENTIAL GUIDE TO THE BUSINESS OF U.S. MOBILE WIRELESS COMMUNICATIONS
Burnham

THE ESSENTIAL GUIDE TO TELECOMMUNICATIONS, SECOND EDITION
Dodd

THE ESSENTIAL GUIDE TO WIRELESS COMMUNICATIONS APPLICATIONS: FROM CELLULAR SYSTEMS TO WAP AND M-COMMERCE
Dornan

THE ESSENTIAL GUIDE TO NETWORKING
Keogh

THE ESSENTIAL GUIDE TO COMPUTER DATA STORAGE: FROM FLOPPY TO DVD
Khurshudov

THE ESSENTIAL GUIDE TO DIGITAL SET-TOP BOXES AND INTERACTIVE TV
O'Driscoll

THE ESSENTIAL GUIDE TO HOME NETWORKING TECHNOLOGIES
O'Driscoll

THE ESSENTIAL GUIDE TO APPLICATION SERVICE PROVIDERS
Toigo

THE ESSENTIAL GUIDE TO KNOWLEDGE MANAGEMENT: E-BUSINESS AND CRM APPLICATIONS
Tiwana

THE ESSENTIAL GUIDE TO MOBILE BUSINESS
Vos & de Klein

THE ESSENTIAL GUIDE TO COMPUTING: THE STORY OF INFORMATION TECHNOLOGY
Walters

THE ESSENTIAL GUIDE TO RF AND WIRELESS
Weisman

The Essential Guide to the Business of U.S. Mobile Wireless Communications

JOHN P. BURNHAM

Prentice Hall PTR, Upper Saddle River, NJ 07458
www.phptr.com

Library of Congress Cataloging-In-Publication Data

Burnham, John P.
The essential guide to the business of U.S. mobile wireless communications / John P. Burnham.
p. cm.
ISBN 0-13-042055-7 (paper)
1. Telecommunications equipment industry—United States. 2. Mobile communication systems—United States. 3. Wireless communication systems—United States. 4. Market surveys—United States. I. Title.

HD9696.T443 U5197 2002
384.5'3—dc21

2001024615

Editorial/production supervision: BooksCraft, Inc., Indianapolis, IN
Acquisitions editor: Bernard Goodwin
Editorial assistant: Michelle Vincenti
Marketing manager: Dan DePasquale
Buyer: Alexis R. Heydt
Cover design director; Jerry Votta
Cover designer: Bruce Kenselaar
Project coordinator: Anne R. Garcia

Prentice-Hall, Inc.
Upper Saddle River, New Jersey 07458

Prentice Hall books are widely used by corporations and government agencies for training, marketing, and resale.

The publisher offers discounts on this book when ordered in bulk quantities. For more information contact:
Corporate Sales Department
Phone: 800-382-3419 Fax: 201-236-7141
E-mail: corpsales@prenhall.com
Or write:
Prentice Hall PTR
Corp. Sales Dept.
One Lake Street
Upper Saddle River, NJ 07458

Printed in the United States of America

10 9 8 7 6 5 4 3 2 1

ISBN 0-13-042055-7

Pearson Education Ltd.
Pearson Education Australia PTY, Ltd.
Pearson Education Singapore, Pte. Ltd.
Pearson Education North Asia Ltd.
Pearson Education Canada, Ltd.
Pearson Educación de Mexico, S.A. de C.V.
Pearson Education—Japan
Pearson Education Malaysia, Pte. Ltd.
Pearson Education, Upper Saddle River, New Jersey

For my daughters,
Miranda & Veronica
and my parents,
Patrica & Walter Dean

Special thanks to Maximo Diego III
for his help with some of the
original research for this project.

Contents

Preface

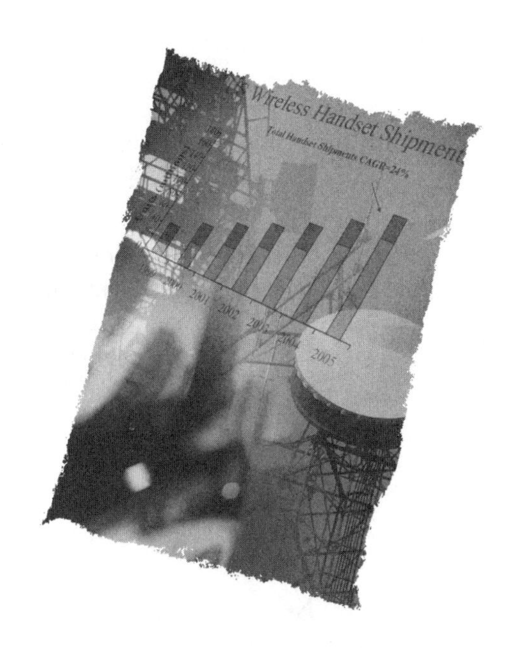

The excitement surrounding the mobile wireless communications industry is palpable and well deserved. At its bedrock, the industry rests on the three fundamental growth drivers of the U.S. economy—digitalization, the speed and power of information processing, and mobility. The mobile wireless communications industry is extremely large and has experienced tremendous growth, with new capabilities being developed and brought to market on an almost daily basis. This growth should be expected to continue for the foreseeable future.

The increase in personal capability enabled by being in constant, untethered communication, as well as having access to virtually any piece of data, from almost anywhere, is unparalleled. Personal capability will continue to expand well into the future, as it is still the early innings of the wireless revolution.

However, there is a tremendous amount of hype surrounding mobile wireless communications. Upon close examination, there remain substantial technical and financial hurdles that must be cleared. While these issues will ultimately be solved, it will take longer than many believe and there will be plenty of casualties along the way.

When an industry has the size and growth trajectory of mobile wireless communications, one would expect the major players to be very profitable. However, due to challenging structural issues, many segments of the industry are unprofitable. The structural issues in mobile wireless communications may well lead to a profitless prosperity, where growth rates are high, but acceptable financial returns are hard to come by.

This book seeks to provide the business professional, investor, or curious soul a clear and concise explanation of the mobile wireless communications industry and its features, supported by facts and figures. It will provide an understanding of the key technologies without being a technical treatise, and will attempt to separate the hype from the realistic business prospects for the major market segments, sometimes with surprising results.

The different segments of the mobile wireless communications market are remarkably different, as are their competitive intensities and business prospects. This text will help to identify the most promising areas and the areas where the risks are high, despite phenomenal growth rates.

As with any endeavor, caution and close study are well advised. This text is designed to help in this effort. As a further note of caution, the mobile wireless communications industry is changing with increasing velocity, and while every effort has been made to ensure this text is accurate and up to date, changes in the industry can overtake the content of this text, and recalibration is strongly advised before taking any action.

No warranties can or will be made with respect to any data, explanations, or opinions expressed herein.

Introduction

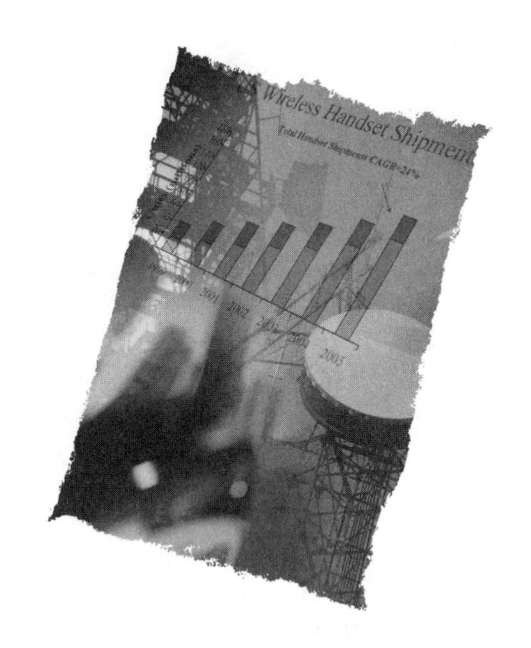

The mobile wireless communications industry is one of the largest, fastest growing industries in the history of the U.S., and unites, as its foundation, the primary growth drivers of the U.S. economy. These growth drivers will continue to propel rapid growth in the industry for the foreseeable future.

The decision to study the mobile wireless communications industry was the result of previous work that attempted to identify the key growth drivers of the U.S. economy. This study found that the major drivers of growth in the U.S. economy could be distilled into three overarching megatrends or themes.

- Digitalization
- Speed and power of information processing
- Mobility

With the primary growth drivers identified, the next step was to ascertain which areas of the economy would benefit the most from these factors. The clearest winner in this environment is mobile wireless communications, which simultaneously harnesses the power of these three principal growth themes. Mobile wireless communications has demonstrated the implicit power of these growth drivers by achieving unprecedented growth, and near ubiquity, in the U.S. market. This growth should be expected to continue well into the future as the industry benefits from these fundamental growth themes.

DIGITALIZATION

Digitalization represents the evolution of communications, media, and electronic devices from analog to digital formats and protocols. Digitalization improves the efficiency and quality of communications, allows devices to communicate with one another, and allows devices to become smaller, cost less, and use fewer moving parts.

There are two separate growth drivers in digitalization—the advent of new digital devices and the substitution of digital devices for existing analog devices. Both areas have grown rapidly.

Digital communications technologies are radically changing the entire economy and the way people live. Digital communications technologies dramatically increase efficiency, throughput, and the number of value-added features that can be enabled.

The advent of the Internet—and the resulting torrent of digital data—has led to exploding demand for bandwidth. This has caused a tectonic shift in the architecture of the telecommunications and cable networks. In a dizzying cycle, the deployment of more bandwidth has led to the development of more bandwidth-hungry interactive digital content such as on-line gaming, video on demand, video conferencing, and content downloading. With digital Third Generation (3G) wireless communications networks, these functions will also be enabled wirelessly. These advances are leading to increased control and interactivity, as more communications, monitoring, and control tasks can be completed through digital wireless networks.

Digitalization leads to overlapping device functionality and new combinations of functionality, enabled by the ease and consistency of communications between devices. This will enable greater control, interactivity, and networking among digital devices, as they are able to speak the same language.

Digitalization will also lead to miniaturization and lower cost devices, as the need for motors and mechanical solutions is reduced by the substitution of chips, or Application Specific Integrated Circuits (ASICs) for moving parts.

Evidence of the impact digitalization has had on our economy can be clearly seen as digital images take over the photography market; CDs and DVDs take over the recorded music and video industries; and books, magazines, newspapers, video, and music are distributed over the Internet.

SPEED AND POWER OF INFORMATION PROCESSING

The speed and power of information processing has increased geometrically and pushed seminal changes through society. The ability to process information has become one of the key abundances of this era, as the plummeting cost of processing a bit of information has approached zero. This has enabled the meteoric growth of the Internet and its resulting applications. One of the most interesting features of this dynamic is that as more information can be processed, more information has been demanded.

The speed and power of information processing has increased along a number of different dimensions that, when leveraged against each other, have led to the geometric expansion in information processing ability.

Microprocessor speed has continued to expand according to Moore's Law, which suggests a doubling of performance every 18 months. ASICs have rapidly improved functionality, and have lowered the cost and reduced the size of electronic devices, as the number of moving parts and complexity has been reduced by providing powerful solutions on a chip.

Fiber-optics technology has substantially increased the available bandwidth for transmission, and Wave Division Multiplexing (WDM) has dramatically increased the throughput of data transmission. This has caused a tremendous change in the telecommunications architecture, as the existing, primarily voice-based circuit-switched network, is converted to a digital data-centric IP-based network. This has also been occurring in the mobile wireless communications industry, as it inexorably marches toward Third Generation broadband packet-switched technologies.

Tremendous progress has also been made in terms of bringing broadband access to the consumer, through DSL, satellite, fixed wireless, cable modems, and powerful digital set-top boxes.

Within the home, wireless networking technologies offering speeds as high as 10 Mbs are being introduced, and hard-wired systems can offer even greater bandwidth.

Wireless networks have substantially increased their bandwidth and throughput by converting from analog to Second Generation digital technologies. These networks will continue to improve, and ultimately achieve Third Generation speeds that are as high as 2 Mbs.

MOBILITY . . . .

Mobility, both within and outside the home, is the third primary growth trend. This dramatic shift has been fueled by digitalization and the increased speed and power of information processing.

Huge increases in subscribership are projected for the mobile wireless communications market, as features are improved through digitalization and miniaturization and as prices for services and hardware drop dramatically. Already, some of the mobile data features, such as wireless Internet access, have come to market, and many people are predicting that virtually all voice communications traffic in the U.S. will be over wireless networks in the next 5–10 years.

Inside the home, customers are beginning to look for wireless networking solutions as they seek to become untethered from phone lines, cables, and electrical outlets. Substantial growth is being projected for wireless networking, wireless modems, and Bluetooth-enabled appliances.

The fastest growth segments of the computing market are mobile products, such as notebook PCs and PDA devices like Palm™ handhelds. Digital wireless handsets will also evolve toward PC functionality. Handsets will be able to communicate with people and machines, perform commercial transactions, process information, and wirelessly control peripheral devices. Portable digital AV devices, such as portable DVD, and MP3 players, will also continue to gain in popularity.

MOBILE WIRELESS COMMUNICATIONS . . . .

If digitalization, the speed and power of information processing, and mobility are accepted as the key growth drivers of the U.S. economy, it is easy to understand why mobile wireless communications has become such a fundamentally exciting and high-growth industry and why it garners the attention that it does. This is particularly true when the Internet is added to the mix.

Digitalization has vastly increased the capacity and throughput of mobile wireless communications from a few kbps to 2G digital, and continuing well into the future with 3G, attaining speeds as high as 2 Mbps. Digitalization has also reduced the terminal size dramatically, shrinking handsets from the size of a brick to something that fits in a shirt pocket. This trend in wireless devices can be expected to continue, as more powerful semiconductors are integrated into the design of these products.

The speed and power of information processing benefits mobile wireless communications, both in terms of the amount of data that can be transmitted and in the amount of data that can be processed locally, at the terminal. 3G wireless speeds are breathtaking, from an historic perspective, as is the processing power of terminals, some of which, like the RIM pager, have on-board Intel 386 processors. Device and network speeds are inextricably linked and following the same rapid growth trajectories, as the speed and power of information processing are extended wirelessly.

Mobility is the primary benefit of mobile wireless communications, and the ability to communicate on the move is the fundamental driver of this trend.

Mobile wireless communications is positioned squarely at the intersection of these megatrends and will continue to benefit handsomely from advances along these dimensions. The industry also has the benefit of being very large. Mobile wireless communications is poised to grow explosively, especially as increased data services are added to the mix. It is unprecedented for an industry of this size to be maintaining growth rates close to 30% per annum. The level of conviction that the primary drivers of growth in the U.S. economy have been properly identified increases substantially when the size and growth rate of this industry are considered. This also enhances the level of confidence that the industry will be able to maintain strong growth, well into the future, as it rides on the coattails of these three primary growth themes.

INDUSTRY STRUCTURE

In beginning this study, it is important to develop a meaningful framework for viewing the industry. The mobile wireless communications industry is very large and very complicated, making it difficult to understand how all the moving parts work together. To this end, it is helpful to develop an appropriate map or model of the value chain and the industry structure, in order to better understand how the industry and its major components operate and interact (see Figure 0.1).

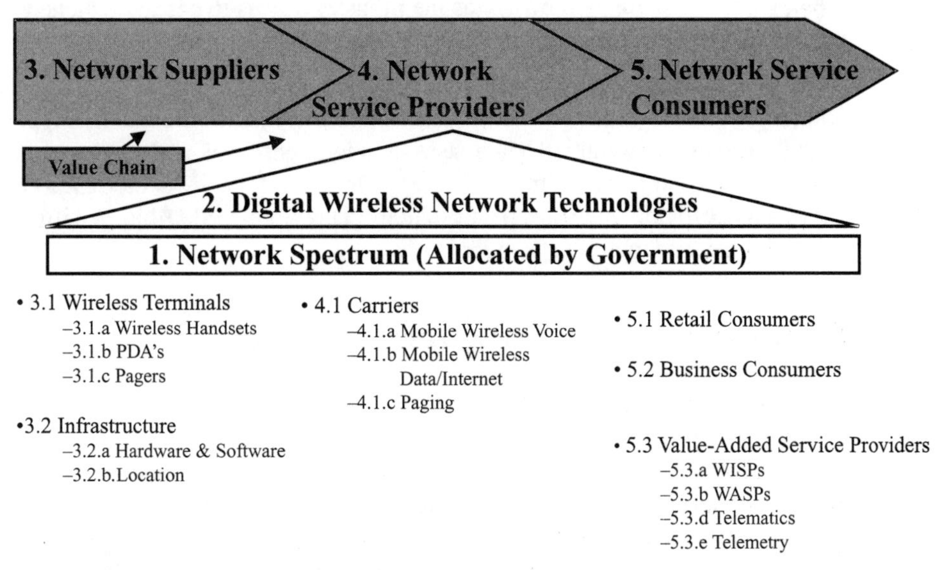

Figure 0.1
U.S. Mobile Wireless Communications Industry Map.

Network Spectrum

Network spectrum is the *sine qua non* of the wireless industry and refers to the ability to transmit signals at a specific frequency and the bandwidth of that frequency. A firm needs to acquire a license from the FCC in order to offer mobile wireless communications in the U.S. Spectrum is the key industry shortage and demand outstrips supply by a wide margin. Thus, spectrum prices have become very, very high.

Digital Wireless Network Technologies

Once adequate spectrum has been secured, a digital wireless network technology must be selected, in order to provision the mobile wireless service.

Digital transmission technologies are the method through which signals are transmitted and received between the handset (terminal) and the base station. One of the key differences between the U.S. and European markets is use of different digital transmission technologies. The three major digital transmission technologies employed in the U.S. are Code Division Multiple Access (CDMA), Time Division Multiple Access (TDMA), and Global System for Mobile Communications (GSM). Additionally, Nextel uses IDEN, a proprietary technology from Motorola that is similar to TDMA and GSM.

These standards will evolve as the industry moves to packet-switched and higher data rate services, known in industry parlance as 2.5G and 3G.

By supporting different technologies, the U.S. market incurs increased cost and complexity, but may also encourage more competition and innovation. Supporting different standards substantially increases development costs and reduces volumes for infrastructure and terminal manufacturers. However, CDMA is thought to be significantly more efficient in using spectrum than TDMA and GSM, thus benefiting the carriers who have adopted this standard.

Paging/Data Network Technologies

Paging/data networks are mobile wireless communications networks that communicate data rather than voice information to the subscriber. They operate under the same principles as voice networks, where the radio signal is transmitted from a base station to a mobile terminal. The major data networks today are Mobitex, DataTac (ARDIS), and the various paging networks throughout the U.S. The vast majority of the U.S. paging networks operate on the FLEX protocol, or its upgrade, ReFLEX.

Some of the major advantages of paging/data networks are extensive nationwide coverage, superior in building reception, ease of use, smaller terminals, and longer battery life.

Network Suppliers

Once spectrum has been obtained and a network technology has been selected, the network needs to be built. This is an extremely expensive undertaking and it can cost billions of dollars to provision a nationwide mobile wireless network.

There are two main components of the mobile wireless network—the infrastructure, which includes base stations and switches, and the terminals, which include wireless handsets, wireless PDAs, and pagers.

The foremost infrastructure suppliers are Ericsson, Nokia, Lucent, Nortel, and Motorola, and the dominant handset suppliers are Nokia, Motorola, and Ericsson. It is very important that the infrastructure and the terminals work well together.

Network Service Providers

Network service providers are the wireless carriers. These are the firms that have secured spectrum, built a network, and provide mobile wireless network services to their subscribers. There are three major classes of network service providers, although the lines between them are beginning to blur. These include the cellular voice and data communications providers like AT&T Wireless, Verizon, and Sprint PCS; data-only network providers like Motient and BellSouth Wireless Data; and the paging service providers like Arch and Metrocall.

Network Service Consumers

Network service consumers are those individuals, or firms, that use wireless services offered by the network service providers. Network consumers can be separated into three distinct segments: retail consumers, business consumers, and value-added service providers.

Retail consumers are individual subscribers who use wireless services primarily for communication, convenience, and safety. Business consumers have two primary motivations for using mobile wireless communications—to increase the productivity of their employees and/or to better address the needs of their customers. Value-added service providers seek to use somebody else's network to provide their own service. Examples of value-added service providers include Wireless Internet Service Providers (WISPs), Wireless Application Service Providers (WASPs), telematics (combination of wireless communications and global positioning in an automobile), and telemetry (machine-to-machine or machine-to-human communication).

Industry Size

When looked at in total, the mobile wireless communications business is enormous. There are currently more than 150 million subscribers to some form of mobile wireless communications. Wireless device sales are projected to top 100 million units in 2001 and industry revenues to be in excess of $50 billion. These figures are all expected to grow substantially over the next several years as subscribership grows, penetration increases, and minutes of use (MOU) expand. Figure 0.2 shows the growth of U.S. wireless device shipments from 1999 and projected to 2005. Figure 0.3 shows wireless revenue in the U.S. for the same time period.

U.S. Focus

Although mobile communications is truly a global industry, and becoming more so on a daily basis, the focus of this text is on the U.S. market. This is due to the fact that the U.S. market is unique in many respects; also it is the world's largest single market.

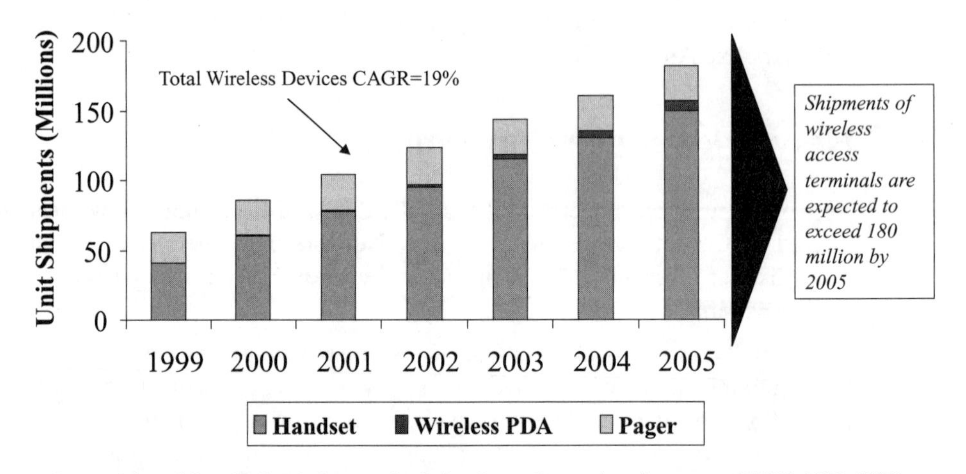

Source: Adapted from IDC and Strategy Analytics, Inc.: a Boston-based company (2000d, 1999, 1998)

Figure 0.2
Unit Shipments of U.S. Mobile Wireless Communications Terminals.

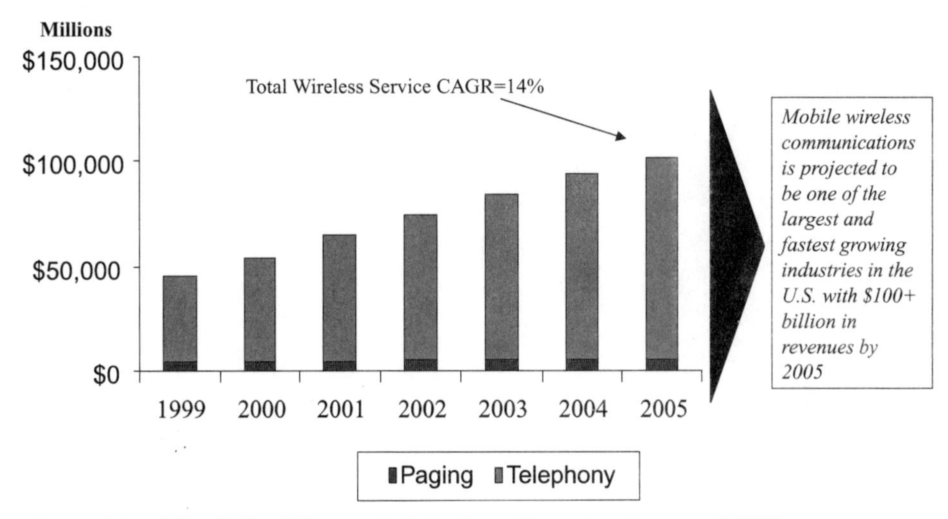

Source: Adapted from IDC and Strategy Analytics, Inc.: a Boston-based company (2000d)

Figure 0.3
U.S. Mobile Wireless Communications Service Revenues.

The most notable difference between the U.S. market and the rest of the world is the array of transmission technologies employed. The U.S. supports a wide variety of standards, while most of Europe is monolithic in approach, using only GSM.

There is much ballyhoo about other parts of the world being ahead of the U.S. in mobile communications. However, the fact remains that the U.S. is, by far, the largest single market for mobile wireless communications in the world.

Areas Not Covered

Two major areas of wireless communication that are not addressed in this text are satellite and fixed wireless communications. Satellite as a mobile wireless communications medium is unlikely to be tremendously relevant in the U.S., due to the reasonably pervasive coverage of terrestrial networks, latency issues, and the high cost of satellite service. Fixed wireless services do not currently support mobility and thus are not included.

1 Mobile Wireless Communications Spectrum

In this chapter...

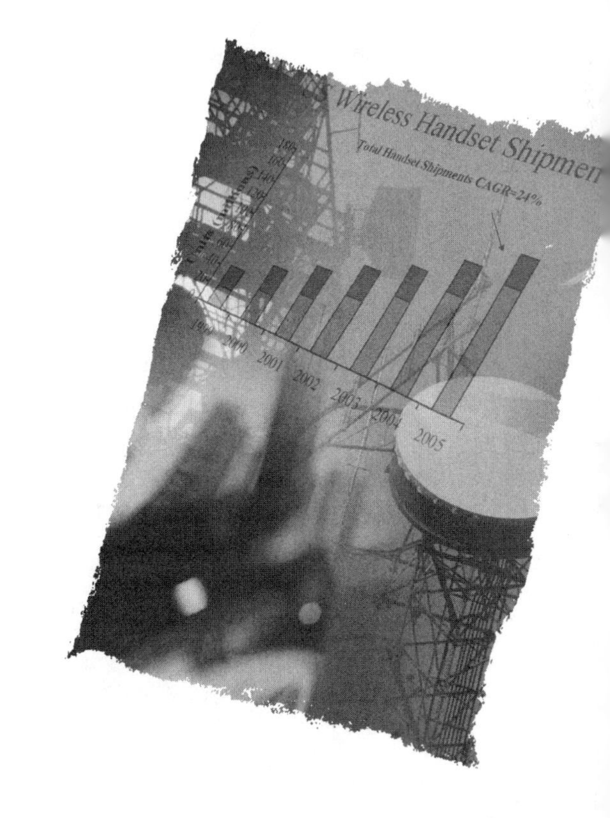

Spectrum, in wireless communications, refers to the ability to transmit signals at a specific frequency, and the bandwidth of that frequency.

The ability to transmit signals at a particular frequency is generally conferred by a national government, in the form of a license. In the United States, regulatory responsibility for radio spectrum is divided between the Federal Communications Commission (FCC) and the National Telecommunications and Information Administration (NTIA). The FCC, which is an independent regulatory agency, administers spectrum for non–Federal government use and the NTIA, which is an operating unit of the Department of Commerce, administers spectrum for federal government use. One of the primary missions of the FCC is to provide advice on technical and policy issues pertaining to spectrum allocation and use in order to most effectively manage the country's radio frequency (RF) spectrum.

Demand for spectrum—from both the private sector and from government agencies—has rapidly outstripped its very limited supply. This is especially true for spectrum that is suitable for mobile wireless communications. Faced with a spectrum shortage, the difficult job of efficiently rationing the available RF spectrum has become one of the FCC's most important tasks.

The major consumers of RF spectrum include wireless communications, television broadcasting, radio broadcasting, and the Department of Defense. The communications portion of the radio spectrum is generally considered to be the range between 9 KHz and 38 GHz. Figure 1.1 outlines the major users of spectrum and their operating frequencies.

Spectrum is the fundamental requirement for entering the mobile wireless communications industry, and is the key shortage in the industry. The available mobile wireless communications spectrum in the U.S. is rapidly becoming filled up, as the number of subscribers and their minutes of use continue to grow rapidly. Additionally, more spectrum will be required in order to accommodate data transfer and offer Third Generation (3G) services.

However, there is not very much spectrum that can be made available for mobile wireless communications in the near future. As such, one would expect the price of spectrum to be very high, and the parties that control it to exert a great deal of power over the market. This is, in fact, the case, and has been played out in dramatic fashion with the astronomical prices paid for licenses in the United Kingdom and German 3G auctions held in the summer of 2000. Additionally, the FCC exerts tremendous influence over the ground rules and profitability of the wireless communications industry in the U.S.

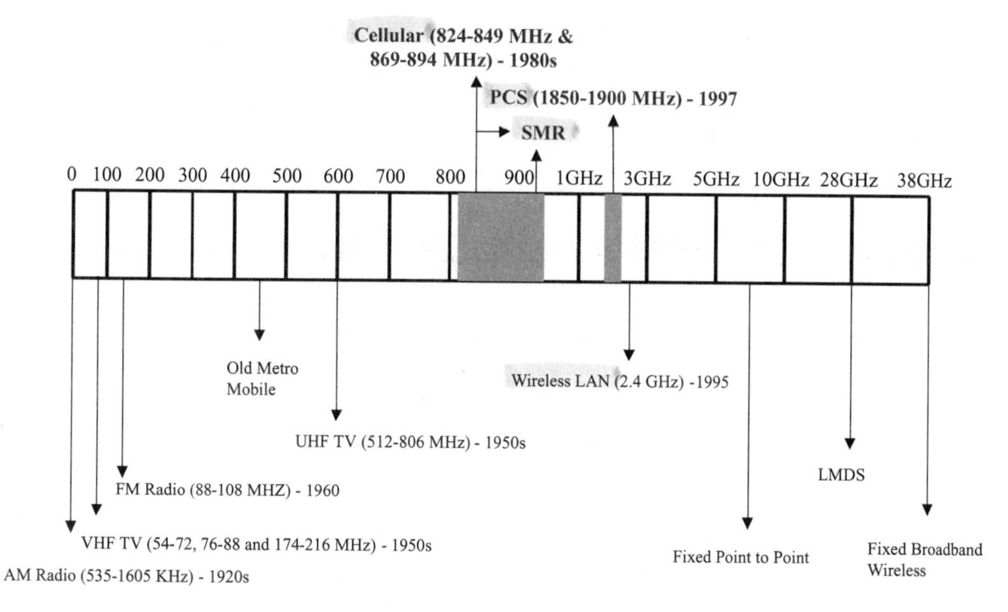

Sources: Telestrategies and First Union

Figure 1.1
U.S. Wireless Frequency Allocation.

PHYSICS OF RADIO COMMUNICATION . . . .

The physics of radio communications involves three interrelated dimensions that are traded off against one another. These dimensions are commonly referred to as the Radio Trinity: frequency, power, and bandwidth.

Frequency is the place on the electromagnetic spectrum where a signal is transmitted or received. At higher frequencies, more information can be transmitted, due to the shorter wavelength. However, the shorter wavelength decreases the travel distance of the signal, and increases the likelihood of interference. Higher frequencies also require larger and more expensive components.

Power is the energy in the wave. When power is increased, travel distance increases and the likelihood of interference is reduced. This is why a 50,000-watt radio station can be heard over a much wider area than a 25,000-watt station. However, using too much power can also pose problems. Hence, engineers spend considerable effort trying to increase the performance of wireless networks at lower power levels.

Bandwidth is the breadth of the frequency used. Bandwidth is analogous to a water pipe, in which a larger diameter allows more water to be moved. Therefore, a 30-MHz wireless license is much more valuable, in terms of capacity and throughput, than a 15- or 10-MHz license.

CELLULAR SPECTRUM ALLOCATION . . . .

Originally, the FCC set aside radio spectrum for cellular service at 824–849 MHz and 869–894 MHz. The FCC divided the country into 306 Metropolitan Service Areas (MSAs) and 428 Rural Service Areas (RSAs). Each MSA and RSA was to have two cellular service providers, in order to foster competition. One 25-MHz license was given to the local wireline company (B Block license), and the second was assigned to a nontelephone company (A Block license). The local wireline company was generally the Regional Bell Operating Company (RBOC), such as Verizon or Bell South. In more rural areas, outside the RBOC's footprint, RSA licenses were given to the rural telephone companies. This enormous government grant accounts, in large part, for the RBOCs' strong position in the mobile wireless communications industry.

Spectrum, like oceanfront property, has generally increased in value, and those firms that have received it for free, or at little cost, have benefited handsomely. This has been particularly true for the network broadcasters (ABC, NBC, and CBS) and for the RBOCs. The rationale for giving these licenses away was to encourage companies to build the mobile wireless communications industry. The main premise of this argument was that service is deployed much more rapidly and thoroughly when spectrum is assigned to a capable service provider at a low cost. The European auctions may provide a good test for this assertion, since the cost of licenses and infrastructure may prove to be so high that reasonable business cases cannot be sustained.

The A Block licensing process was slightly more complicated. The first 30 markets were awarded based on comparative hearings. However, this process proved to be extremely unwieldy and was abandoned in favor of a lottery system. Many of the players in the lottery system reduced their risk by forming joint ventures to secure partial ownership of licenses.

The result of this process was very fragmented ownership of A Block licenses. This paved the way for A Block consolidators like McCaw (subsequently acquired by AT&T) to establish large A Block spectrum positions. Additionally, the RBOCs were allowed to purchase out-of-region A Block licenses from the lottery winners. Several examples of this were SBC Communications, Inc., purchasing licenses in Boston, Washington, and Chicago, and PacTel purchasing licenses in Atlanta, Ohio, and Michigan. This cellular duopoly was allowed to exist for many years. However, there were many complaints that this system did not foster sufficient competition or innovation.

PCS SPECTRUM ALLOCATION . . . .

Due to consumer complaints and the need for more mobile wireless communications capacity, Congress directed the FCC to make more spectrum available through a competitive auctioning process as part of the 1993 Omnibus Budget Reconciliation Act. Pursuant to this, the FCC announced plans to auction off 120 MHz of the 1850–1990 MHz personal communications services (PCS) spectrum. The PCS spectrum was divided into six segments to be auctioned: the A, B, C, D, E, and F blocks.

The A and B blocks were for 30 MHz each in the 51 Major Trading Areas (MTAs). MTAs are regions that include multiple cities or states.

The C (30-MHz) and D through F (10-MHz) blocks provided coverage for the 493 Basic Trading Areas (BTAs). BTAs are regions that include only one metropolitan area.

The primary goals of the PCS auctions were to increase competition in the industry and to raise money for the Treasury. Both of these goals were achieved. To foster further competition, the incumbent cellular providers were prohibited from bidding on the new PCS licenses in that same market. This substantially increased competition, and resulted in most areas having two cellular providers and as many as six new PCS suppliers. The government also raised more than $20 billion in proceeds.

The auction for the A and B blocks began in December 1994 and lasted 112 rounds, concluding on March 13, 1995. The auction raised $7.7 billion in proceeds, and the major winners were Sprint PCS and AT&T Wireless.

The C Block auction, which was limited to qualified small businesses by order of Congress, began in December 1995 and ended May 6, 1996, after 184 rounds of bidding. This auction raised about $10 billion, which was well above expectations. This may have been due to low-cost government financing (6.5–7.0% annual interest rate) that was offered to these entrepreneurs. The biggest winner in this auction was Nextwave Personal Communications.

The D, E, and F blocks (the F Block was restricted, like the C Block, to qualified small businesses) were auctioned from August 1996 until January 14, 1997. After 275 rounds, $2.5 billion was paid for these licenses. Sprint PCS and AT&T Wireless came away with the most licenses in the D and E block auctions, and Nextwave was the big winner in the F Block auction. Through the C and F block auctions, Nextwave was able to establish virtually nationwide license coverage. Table 1.1 shows a summary of PCS license auctions in the U.S.

Table 1.2 Summary of Previous PCS License Auctions in the U.S.

Band	Date Completed	Price (millions)	POPs (millions)	Bandwidth (MHz)	Price per POP	Price per POP per MHz
A/B	Mar-95	$7,736	276	60	$28	$0.47
C	May-96	$9,967	276	30	$36	$1.20
D/E/F	Jan-97	$2,523	276	30	$9	$0.30
Total		**$20,226**	**276**	**120**	**$73**	**$0.61**

Source: Salomon Smith Barney, 2000

Note: POP refers to a single individual in a region. Note that in wireline telephony POP refers to a "Point of Presence" or a physical location in an area.

SPECIALIZED MOBILE RADIO SPECTRUM ALLOCATION

Specialized Mobile Radio (SMR) represents 26.5 MHz of spectrum located in the 800- and 900-MHz bands (see Figure 1.2). SMR spectrum was allocated by the FCC in a piecemeal fashion on a first-come first-served, tower-by-tower basis. Subsequently, much of this spectrum was acquired by Nextel. Nextel was able to construct a nationwide network with this spectrum, with an average nationwide bandwidth of 19 MHz in 2000.

UNLICENSED SPECTRUM

Please note that there are also some unlicensed frequencies where other wireless devices can operate. These devices include cordless phones, car door remote controls, garage door remotes, and many others. The unlicensed 2.4-GHz frequency has become very widely used and is rapidly becoming overcrowded with consumer devices.

With the advent of Bluetooth, more spectrum will be needed to accommodate wireless networking of devices. Bluetooth uses tiny radio transceivers that operate in the unlicensed 2.4-GHz band, to transmit voice and data at rates up to 721 kbps. Bluetooth communicates within a 10-meter perimeter and does not require line of sight to establish a connection. If Bluetooth takes off as predicted, the 2.4-GHz band will become even more crowded, and the band will probably have to be expanded.

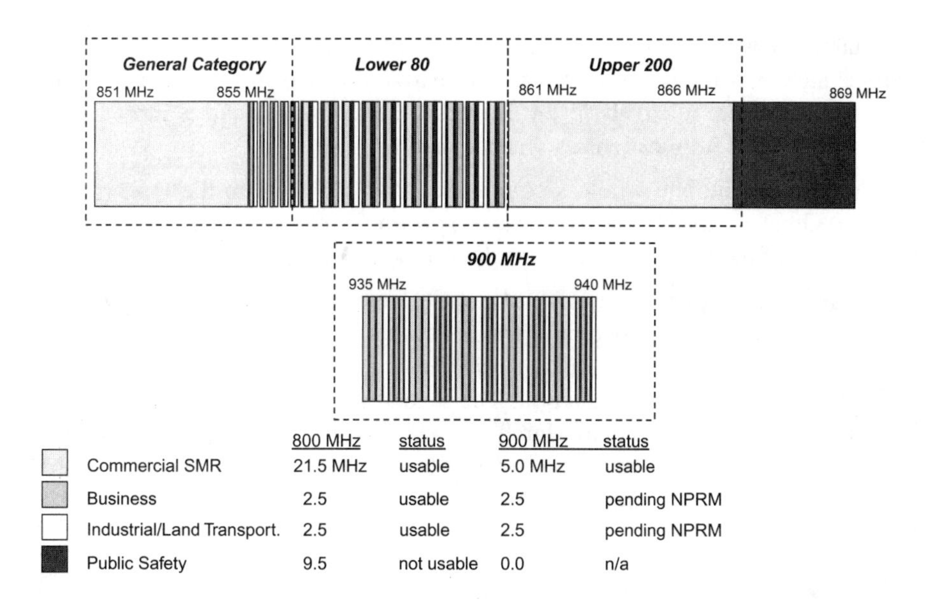

Note: Above chart shows "send" channels—each is paired with a receive channel.
Use of 900-MHz Business/Industrial and Land Transportation channels is being considered in a Notice of Proposed Rulemaking (NPRM).
Source: FCC

Figure 1.2
SMR Frequency Allocations at 800 and 900 MHz.

TOTAL SPECTRUM AND SPECTRUM CAPS . . . .

The total amount of spectrum devoted to mobile wireless telephony in the U.S., as of January 1, 2001, was 50 MHz of cellular, 120 MHz of PCS, and 26.5 MHz of SMR, for a total of 196.5 MHz. In order to maintain competition, the FCC has limited the total amount of spectrum that any one carrier can own in any single market. The limits as of January 1, 2001, were 45 MHz in a metropolitan market and 55 MHz in a rural market. As a result of these caps, there are a minimum of five competitors in metropolitan markets and four competitors in rural markets. In January 2001, the FCC was accepting input on whether these caps should be raised in the future.

SPECTRUM AUCTIONS AND REAUCTIONS . . . .

The ability of the government to make more spectrum available is one of the key features of the industry. The government can and will make more spectrum available. They have both a public policy and financial motive to do so. This is a potent combination in a political process, and should not be underestimated.

Making new spectrum available is generally good for consumers, since prices usually come down and new services are made available. However, it generally makes the existing spectrum relatively less valuable as the industry becomes more competitive. This has broad implications for carriers and for investors.

The existing mobile wireless communications spectrum in the U.S. is rapidly filling up. Greater capacity, in the form of spectrum, will be required to enable the future growth of the industry and the evolution to 3G wireless data services.

To make more spectrum available the FCC has scheduled two auctions. The first is the reauction of 1.9-GHz spectrum, which began on December 12, 2000, and ended on January 26, 2001. This spectrum is particularly valuable since it can be used almost immediately after winning the auction. The second is the auction of 700-MHz spectrum that was scheduled to begin March 6, 2001. There are many issues surrounding this spectrum. These issues have caused the 700-MHz auction to be further delayed by the FCC, until at least September 12, 2001.

1.9-GHz PCS Spectrum Reauction

The 1.9-GHz spectrum reauction refers to spectrum that was successfully bid on in the original PCS auctions, but where the winning firm was never able to deploy service, usually as a result of bankruptcy. The FCC had maintained the right to take back this spectrum in the original auction if service was not deployed according to certain schedules. The largest firm unable to provide service was NextWave Personal Communications, which had winning bids on a collection of licenses that would have enabled it to offer nationwide service. For several years, NextWave tried on a number of legal and legislative fronts to block the surrender of its licenses. However, the FCC prevailed and was able to reauction the licenses. The FCC began reauctioning this spectrum on December 12, 2000.

The 1.9-GHz spectrum available for reauction encompasses Metropolitan Trading Areas (MTAs) throughout the country, from Honolulu to New York City, in bandwidths varying from 10 to 30 MHz. This spectrum represents the only clean spectrum that will be able to be auctioned for the foreseeable future. It is anticipated that the major carriers will bid aggressively, viewing this auction as an opportunity to shore up areas where they are deficient in spectrum or coverage. This auction was successfully completed on January 26, 2001, after 101 rounds of bidding had been completed. The total bid reached $16.857 billion with Verizon winning the most licenses. This amount translates to $4.08 per MHz per POP, and is comparable to the United Kingdom and German auctions held in the summer of 2000, which realized about $4.08 and $4.66 per MHz per POP, respectively. The standard in the industry is to quote spectrum acquisition costs in terms of cost per MHz per POP, which measures the cost of bandwidth per available customer. Table 1.2 shows the reauction results for the top 15 markets.

Table 1.2 1.9-GHZ Reauction Results in the Top 15 Markets by Bid.

Market	High Bidder	Population (mil)	Net Price ($mil)	Price per MHz per POP
New York, NY	Cellco Partnership	18.1	2,057	$11.40
New York, NY	Cellco Partnership	18.1	2,038	$11.29
New York, NY	Alaska Native	18.1	1,484	$8.22
Los Angeles, CA	Cellco Partnership	14.5	514	$3.53
Chicago, IL	Cellco Partnership	8.2	495	$6.05
Los Angeles, CA	Alaska Native	14.5	435	$2.99
Los Angeles, CA	Salmon PCS	14.5	409	$2.81
San Francisco, CA	Cellco Partnership	6.4	399	$6.21
Atlanta, GA	Salmon PCS	3.2	322	$10.07
Philadelphia, PA	Cellco Partnership	5.9	277	$4.70
Washington, DC	Cellco Partnership	4.1	217	$5.26
Dallas, TX	Salmon PCS	4.3	214	$4.94
Boston, MA	Cellco Partnership	4.1	212	$5.13
Boston, MA	Cellco Partnership	4.1	192	$4.63
Washington, DC	DCC PCS	4.1	172	$4.18
Total Auction		**174**	**$16,857**	**$4.08**

Cellco was a Verizon operative
Alaska Native was an AT&T Wireless operative
Salmon PCS was a Cingular operative

Source: FCC, 2001

To put current spectrum costs in perspective, a simple back-of-the-envelope calculation is helpful.

$4.20 per MHz per POP × 30 MHz	$126
Divided by wireless penetration	40%
License cost per active POP	$315
Divided by proportional share based on minimum number of active carriers	25%
Spectrum cost per proportional active subscriber	$1,260

Taking spectrum cost per subscriber and dividing it by a monthly Average Revenue per User (ARPU) of $45 per month, it would take 28 months for a carrier just to recoup its spectrum acquisition costs if they were able to attract 25% of the existing market's wireless subscribers as net new customers. Calculating the cost of this spectrum based on incremental new customers, or added services, which is the appropriate measure, makes the business case even more challenging. Admittedly, the calculation carriers must make is far more complicated; however, the back-of-the-envelope numbers point out the substantial risks associated with acquiring spectrum at these prices.

700-MHz Spectrum Allocation

In 2001, the FCC is planning to make more spectrum available for mobile wireless communications by auctioning licenses in the 700-MHz spectrum. The 700-MHz spectrum has particularly good characteristics for wireless communications because of its propagation qualities, which allow the waves to travel long distances. This spectrum is also suitable for 3G since the bands are at least 10 MHz wide, the minimum requirement for Wideband CDMA (W-CDMA), and the structure of the auction allows nationwide licenses to be acquired. Therefore, there is very substantial interest in acquiring this spectrum.

However, this spectrum is currently occupied by television broadcast channels 60–69 (see Figure 1.3). This spectrum is due to be returned to the FCC, by the broad-

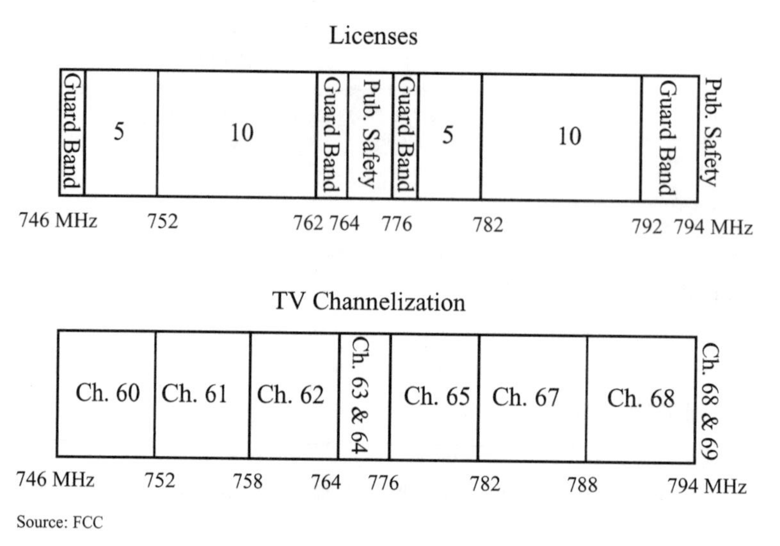

Figure 1.3
U.S. 700-MHz Band Plan.

casters, when the transition to digital broadcasting is complete. The digital broadcasting transition is defined as being complete when 85% of customers have the required hardware to receive the digital television signal, and the digital broadcast signal is being transmitted. The current deadline for this process to be completed is 2006.

To enable digital broadcasting and HDTV, the FCC gave spectrum for digital television to the terrestrial television broadcasters. The broadcasters were to simultaneously transmit in digital and analog until the transition was complete. This was because not everyone would be able to go out and buy a new digital TV the moment the switch to digital was made. Once the transition is technically complete, the broadcasters are required to return the analog spectrum to the FCC. However, this is unlikely to occur before the 2006 deadline, at the earliest.

Giving this much spectrum to the broadcasters for free was widely criticized at the time as a huge and unnecessary "give-away." Bolstering this argument was the fact that most people receive their TV signal from cable companies or satellite services like DirectTV, rather than from terrestrial broadcasters.

Unfortunately, the evolution to digital television has been fraught with delay and disagreement, and many issues still need to be resolved. It will likely be many years until the transition is technically complete, and probably past the 2006 deadline.

The 700-MHz auction has been delayed several times, due to politics and the need to clear the spectrum of its previous users. The carriers are justifiably concerned about bidding for licenses that need to be cleared. Winning bids risk being held hostage by the incumbent broadcasters, like Paxson Communications and Sinclair Broadcasting, who have no intention of vacating the spectrum before they have to, or without securing a large price. This is a huge issue, since the regulations are written so that the broadcasters could technically hold on to this spectrum for a very long time—at least until 2006, five years after the proposed auction. If the FCC decides to step in and take back the spectrum, it will likely face long and expensive court battles, with no guarantee of prevailing.

Due to the uncertainty surrounding the ability, and ultimate cost, to use this spectrum, the government is unlikely to realize top dollar for these assets in an auction. Therefore, it is in all parties' best interest that a clearing agreement or mechanism be in place prior to the auction. There are several proposals for providing a clearing mechanism for this spectrum simultaneously with the spectrum auction; however, none have been finalized or widely agreed upon.

Despite these difficulties, the government is planning on holding this auction on September 12, 2001, having postponed the auction from September 2000 and March 6, 2001, as a result of the aforementioned issues.

700-MHz Spectrum Auction Rules

The 700-MHz auction is set up for 2 licenses (10 and 20 MHz) in six regions (Northeast, Mid-Atlantic, Southeast, Great Lakes, Central/Mountain, and Pacific) for a total of 12 licenses. The 20-MHz license is for the 752–762 MHz and 782–792 MHz bands, while the 10-MHz license is for the 747–752 MHz and 777–782 MHz bands. In this auction interested parties will be allowed to bid, on an all-or-nothing basis, for national licenses. Figure 1.4 details the rules for the 700-MHz auctions. Table 1.3 shows the licenses to be auctioned.

- Bidders may place bids on individual licenses and may also place "all-or-nothing" bids on up to 12 packages of licenses of their own design. "All-or-nothing" package bids ensure the bidder does not face the risk of winning only some of the desired licenses in a package. For example, a company may place a package bid for six licenses to create a national footprint or place a package bid for two licenses in a particular region.

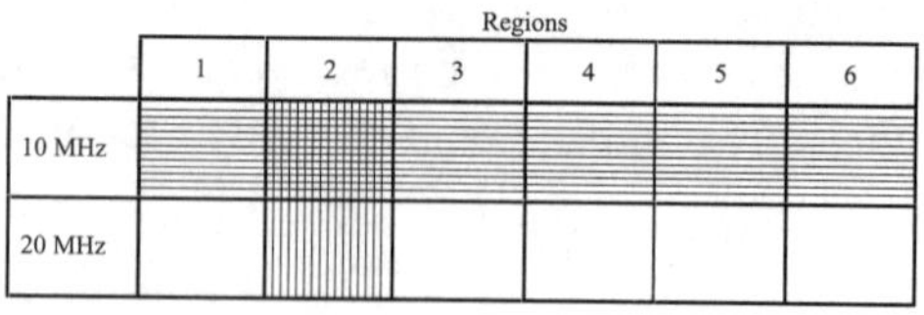

- Bidding will continue until two consecutive rounds have occurred in which no new bids are accepted.
- Winning bids are the set of bids on individual licenses and packages that maximize revenue when the auction closes.

Figure 1.4
Rules for Upcoming 700-MHz Spectrum Auctions.

Table 1.3 700-MHz Licenses to Be Auctioned.

Description	Population (1990)	License Bandwidth (MHz)	Upfront Payment	Minimum Opening Bid
Northeast	41,567,654	10	$14,000,000	$40,000,000
Mid-Atlantic	42,547,218	10	$14,000,000	$40,000,000
Southeast	44,516,919	10	$14,000,000	$40,000,000
Great Lakes	41,560,906	10	$14,000,000	$40,000,000
Central / Mountain	40,926,284	10	$14,000,000	$40,000,000
Pacific	41,427,686	10	$14,000,000	$40,000,000
Subtotal			*$84,000,000*	*$240,000,000*

Table 1.3 700-MHz Licenses to Be Auctioned (continued).

Description	Population (1990)	License Bandwidth (MHz)	Upfront Payment	Minimum Opening Bid
Northeast	41,567,654	20	$28,000,000	$80,000,000
Mid-Atlantic	42,547,218	20	$28,000,000	$80,000,000
Southeast	44,516,919	20	$28,000,000	$80,000,000
Great Lakes	41,560,906	20	$28,000,000	$80,000,000
Central / Mountain	40,926,284	20	$28,000,000	$80,000,000
Pacific	41,427,686	20	$28,000,000	$80,000,000
Subtotal			*$168,000,000*	*$480,000,000*
Total			**$252,000,000**	**$720,000,000**

Source: FCC

WORLDWIDE 3G SPECTRUM BANDS . . . .

On October 13, 2000, President Clinton directed the Secretary of Commerce to work with the FCC to develop a plan to select spectrum for 3G wireless systems by October 20, 2000. They were then instructed to issue an interim report by November 15, 2000. The report was to cover current spectrum uses and the potential for reallocation of the spectrum bands identified for worldwide 3G services at the World Radio Conference-2000 (WRC 2000). The spectrum selection is then to be finalized by July 2001, in conjunction with the NTIA, and the auction is to be completed by September 30, 2002.

The three spectrum bands identified at WRC-2000 were 806–960 MHz, 1710–1885 MHz, and 2500–2690 MHz. In the U.S., these bands are heavily occupied. The primary occupants of this spectrum on January 1, 2001, were analog cellular phone carriers, the Department of Defense, fixed wireless services, satellite broadcasters, school systems, and private video conferencing.

Because these bands are crowded, as is depicted in Figure 1.5, it is unlikely that they can be effectively cleared in the medium term. As a result the U.S. will, most likely, not be consistent with the worldwide 3G spectrum bands and President Clinton's order will not be followed.

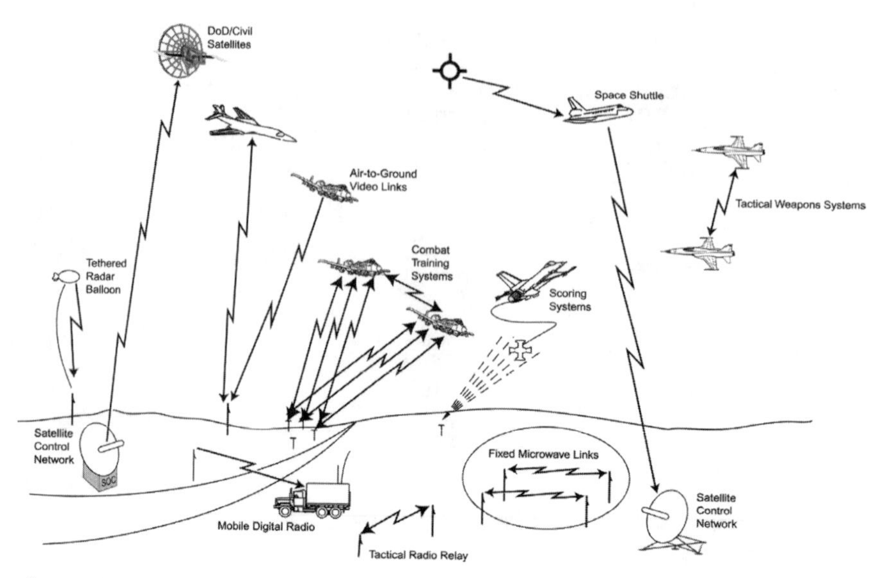

Source: NTIA, U.S. Department of Commerce

Figure 1.5
Functions Supported in 1710–1850 MHz Bands.

EUROPEAN 3G SPECTRUM AUCTIONS

In the summer of 2000, many of the major European governments auctioned the spectrum reserved for 3G mobile wireless communications. The amounts paid for these licenses were truly spectacular and were well above any preauction forecasts.

UK 3G Auction Summary

The British government held its 3G auctions in the summer of 2000. The results of these auctions exceeded even the most optimistic forecasts, generating roughly $35 billion (roughly seven times more than expected) for the government. See Table 1.4 for a list of the United Kingdom 3G auction winners.

There were five licenses auctioned for the United Kingdom, and Vodafone placed the largest single bid—more than $9 billion—for the license with the broadest capacity. These licenses are for 20 years and will come into use in 2002.

Service providers such as Vodafone considered the licenses critical to their ability to offer consumers new high-bandwidth services. However, analysts and consumers fear that the high license prices will result in consumer price increases or

Table 1.4 UK 3G Auction Winners.

Winners	License	Price (millions)	POPs (millions)	Bandwidth (MHz)	USD/POP	Price per POP per MHz
TIW	A	$6,809.4	59.0	35	$115	$3.30
Vodafone	B	$9,262.1	59.0	35	$157	$4.49
BT3G	C	$6,258.7	59.0	25	$106	$4.24
One2One	D	$6,217.6	59.0	25	$105	$4.22
Orange	E	$6,359.5	59.0	25	$108	$4.31
Total		**$34,907.4**	**59.0**	**145**	**$592**	**$4.08**

Source: Salomon Smith Barney, 2000

unsustainable business models. The concern is that the required investments and cost increases will subsequently restrict the market for mobile e-commerce. A cursory look at the cost of the spectrum and the cost of installing the required network and terminals calls into question the ability to build a profitable business with these licenses.

Germany 3G Auction Summary

Germany received $45.9 billion from the auction of six 3G licenses—again, much higher than the most optimistic forecasts. The largest single bid was $7.7 billion from T-Mobil, a subsidiary of Deutsche Telekom. The total proceeds from the auction exceeded the UK auction total by more than $10 billion, although the largest single bid for a German license was less than the highest single bid for a UK license although the bandwidth of the German license was smaller (20 vs. 35 MHz). See Table 1.5 for a list of the German 3G auction winners.

Germany is considered a critical market for service providers seeking to offer pan-European service, and is thought to have high potential for growth in mobile wireless communications given its large, affluent population and lower mobile phone usage than neighboring countries.

The spectacular bids for these licenses continue to concern investors who worry about the billions being spent on licenses and the additional billions that will be spent to build 3G networks. At these levels even service providers have become squeamish. In fact, Hutchinson Whampoa dropped out of the E-Plus Hutchinson consortium citing the high cost of the license.

Table 1.5 German 3G Auction Winners.

Winners	Price (millions)	POPs (millions)	Bandwidth (MHz)	USD/POP	Price per POP per MHz
E-Plus Hutchinson	$7,620.0	82.0	20	$92	$4.65
Group 3G	$7,630.0	82.0	20	$93	$4.65
Mannesmann Mobilfunk	$7,650.0	82.0	20	$93	$4.66
Mobilcom Multimedia	$7,600.0	82.0	20	$93	$4.63
T-Mobil	$7,690.0	82.0	20	$94	$4.69
Viag Interkon	$7,670.0	82.0	20	$94	$4.68
Total	**$45,860.0**	**82.0**	**120**	**$559**	**$4.66**

Source: Total Telecom, 2000

Rest of Europe 3G Auction Summary

Following the huge cost of the UK and German auctions, prices for 3G licenses for the rest of Europe moderated substantially. This was due to investor concern—reflected in falling stock prices and credit ratings in the wake of the German and UK auctions—and more tempered levels of enthusiasm regarding the prospects for 3G services. Figure 1.6 shows the dramatic drop in 3G auction prices following the German and UK auctions.

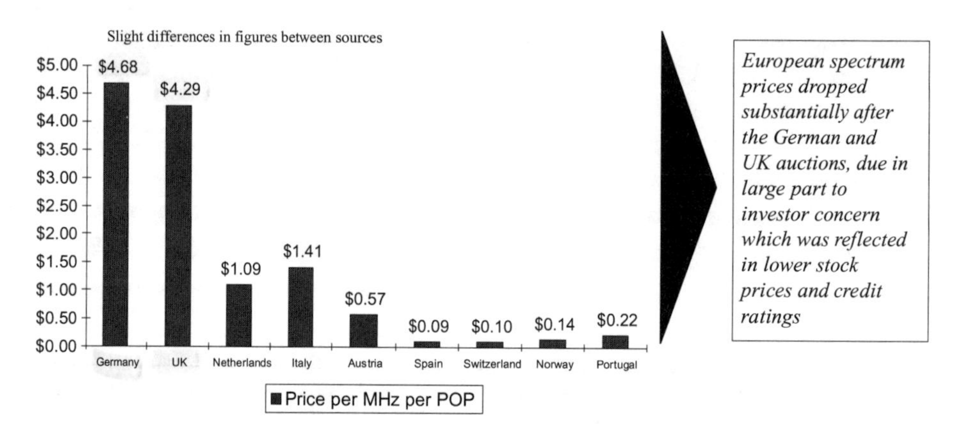

Source: European Telecommunications Authorities and Lehman Brothers, 2000

Figure 1.6
European 3G Auction Results.

OUTLOOK

The spectrum situation in the U.S. is very clouded. Spectrum is the most critical shortage in the wireless communications industry, and there is a severely limited supply. Additional spectrum that is suitable for mobile wireless communications is already heavily occupied, and the ability to clear this spectrum or relocate current occupants is limited and likely to be extremely expensive. At this point, it appears unlikely that the WRC 3G bands will be available in the U.S. market for 3G services.

The lack of new available spectrum is one of the primary reasons that the 1.9-GHz auctions were so well bid. Additionally, the 1.9-GHz spectrum is available and usable virtually immediately, since there are no current occupants and no new equipment needs to be developed to support it. This auction commanded prices similar to the 2000 UK and German 3G auctions; and prices were even higher in New York City and other key markets. At the time, UK and German prices were considered exorbitant, leading to substantial investor concern and falling stock prices and credit ratings for the winning bidders.

The 700-MHz auction is more problematic because a clearing mechanism needs to be devised in order to make the spectrum more valuable. However, since this will be the last auction for the foreseeable future, the spectrum characteristics are ideal for mobile wireless communications, the spectrum band is wide enough to support W-CDMA, and nationwide licenses can be secured, there should be substantial interest in these licenses.

As spectrum is the key shortage in the mobile wireless communications industry, spectrum prices are behaving as one would expect. Since the demand for spectrum outstrips supply, the spectrum holders and rule makers will continue to exert substantial influence over the industry. In the absence of regulation, one would expect spectrum to create a substantial barrier to entry for the industry, and in many respects it does. However, FCC spectrum caps ensure that there is fierce competition in the industry, with a minimum of five competitors in metro markets and four competitors in rural markets. The FCC, in 2001, was beginning to review its spectrum caps. This may bode well for industry incumbents and reduce the competitive intensity of the industry over time.

2 Digital Wireless Network Technologies

In this chapter...

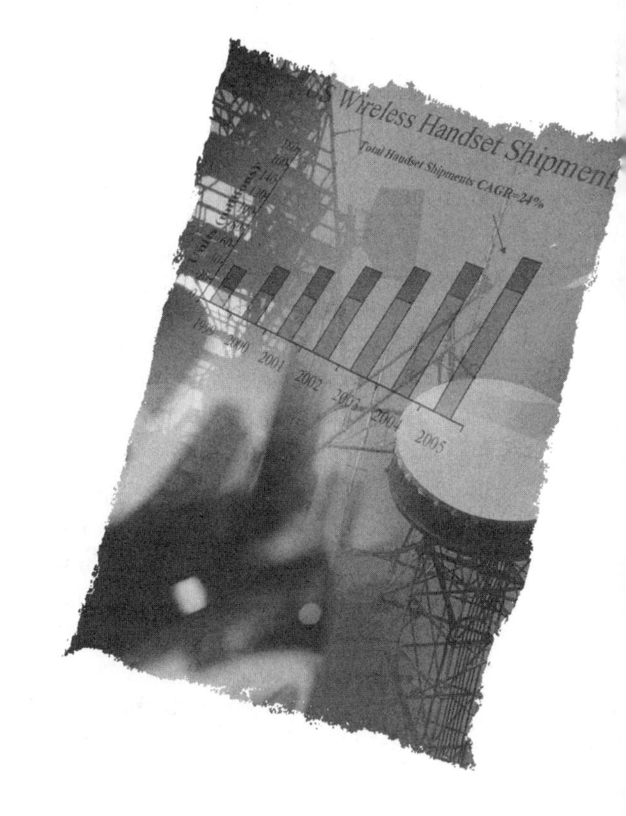

Digital transmission technologies are the method through which radio signals are transmitted and received between the handset (terminal) and the base station. The use of different digital transmission technologies is one of the key differences between the U.S. and European markets. The three major digital transmission technologies employed in the U.S. are Code Division Multiple Access (CDMA), Time Division Multiple Access (TDMA), and Global System for Mobile Communications (GSM). Additionally, Nextel uses IDEN, a proprietary technology from Motorola that is similar to TDMA and GSM.

Europe, and much of Asia, employ a single transmission technology standard, GSM. Using a single standard allows customers to roam throughout Europe with the same handset. The use of a single standard is credited for the tremendous growth and penetration of mobile wireless communications in Europe. In 2000, many European countries had more than 50% market penetration of mobile wireless telephony, with some of the Scandinavian countries in excess of 70%, versus below 40% penetration in the U.S.

By supporting different technologies, the U.S. market incurs increased cost and complexity, but may also encourage greater competition and innovation. Supporting different standards also means that development costs are substantially increased and unit volumes for infrastructure and terminal manufacturers are reduced. However, CDMA is thought to be significantly more efficient in using spectrum and to offer an easier migration path to 3G than TDMA or GSM, thus benefiting the carriers who adopted this standard.

Digital wireless technologies have been adopted due to their clear superiority over analog transmission technologies. The most important benefit of digital transmission is the substantial improvement in spectrum use. Spectrum is the key scarcity in the wireless industry and is tremendously valuable, as evidenced by the 3G auctions in Europe and the 1.9-GHz reauction in the U.S. (see Chapter 1). Digital technologies also increase capital efficiency, lower network operating costs, and offer higher quality service. Additionally, digital transmission technologies enable digital calling features, such as caller ID.

As mobile wireless communications evolve, there is substantial market interest in agreeing on a global standard for 3G, or third-generation, technologies. This would enable global roaming on a single device and substantially reduce research and development costs. However, reaching consensus on a global standard may be unlikely, due to the extremely competitive nature of the industry and national governments. This is particularly true for the U.S. where there is support for at least two distinct technology road maps to 3G.

In the event a common standard cannot be agreed upon, achieving full interoperability between standards and devices would be the next most important priority. Enabling global roaming through interoperability is viewed as more likely than agreeing on a single standard. However, achieving global interoperability could take many years to realize, particularly in light of the seeming inability of the U.S. to clear additional 3G spectrum in the bands outlined in World Radio Conference-2000 (WRC-2000; see Chapter 1).

TYPES OF MOBILE WIRELESS TELEPHONE SERVICE

Before delving too deeply into the different digital transmission technologies it is helpful to clear up a little bit of terminology regarding wireless telephony. Although mobile wireless telephony is generally referred to as cellular ("call me on my cell phone"), this is not entirely accurate and the subtle difference in terminology should be understood.

The three types of wireless telephone services currently offered in the U.S. are all commonly referred to as cellular. This is because the fundamental architecture of the networks remains cellular—each base station covers a particular geographic cell, hence cellular. This is a different meaning than cellular wireless telephone service, which refers to the 800-MHz spectrum band. This can be confusing, especially since Sprint PCS has run an ad campaign suggesting cellular service is inferior to PCS service, even though their network has a cellular architecture. Network architecture is discussed more thoroughly in Chapter 3.

The three types of mobile wireless telephony offered in the U.S. are:

- Cellular
- Specialized Mobile Radio (SMR)
- Personal Communications Services (PCS)

Cellular and SMR (primarily Nextel) use radio spectrum in the 800–1000-MHz band. PCS operates at higher frequencies, in the 1850–1990-MHz band. Figure 2.1 shows frequency allocation in the U.S. At higher frequencies, the signal can carry more information, but is weaker and does not travel as far. By transmitting at the higher PCS frequencies, PCS operators can require up to four times as many cell sites as a cellular provider to cover the same amount of terrain. This can be irrelevant in urban areas, where capacity more than coverage determines the number of cells used. However, PCS build-outs in rural areas can be very expensive. It is important to remember that the key reason PCS was adopted was the availability of the spectrum, not its technological superiority. In fact, for mobile wireless applications, lower frequencies are generally preferred. This is an example of technology being developed to alleviate the key industry shortage, spectrum.

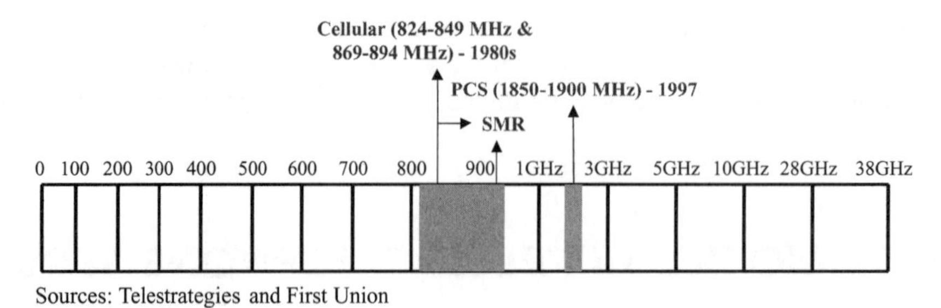

Figure 2.1
U.S. Wireless Frequency Allocation.

DIGITAL TRANSMISSION TECHNOLOGIES . . . .

The different types of service (Cellular, SMR, PCS) are capable of supporting the different air interfaces or digital transmission technologies (i.e., TDMA, CDMA, GSM). The spectrum is agnostic, in this sense, and the carrier will determine which technology to use, once it has secured the spectrum on which to transmit.

Today's circuit-switched, digital-transmission technologies (TDMA, CDMA, GSM) are considered Second-Generation (2G) technologies. First Generation (1G) refers to analog technologies, which use spectrum or bandwidth very inefficiently. Third Generation, or 3G technologies as they are commonly referred to, are broadband, packet-switched technologies.

1G = Analog
2G = Digital Circuit Switched
3G = Broadband Packet Switched

Despite the popularity of digital transmission, from both a customer and a carrier perspective, there was still a large amount of analog transmission in service in the U.S. in 2000. Figure 2.2 shows the breakdown of subscribers by technology. This is particularly true in the more rural areas where the competitive and financial circumstances have not justified the cost of upgrading to digital transmission. Additionally, the FCC requires the original analog companies to continue to maintain analog service on a portion of their spectrum to accommodate existing analog subscribers. In order to be functional in large parts of the country, U.S. digital handsets have had to incorporate an analog transceiver. This is the reason handsets in the U.S. tend to be much larger than phones in the rest of the world.

All digital wireless technologies use a process called multiplexing to carry more calls than analog on the same amount of spectrum. Multiplexing enables signals from

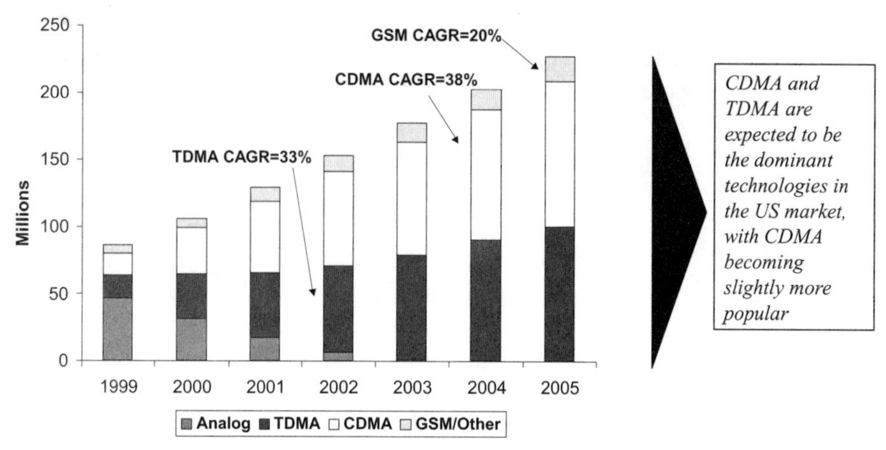

Source: Strategy Analytics, Inc.: a Boston-based company (2000d)

Figure 2.2
U.S. Wireless Subscribers by Technology.

one or more devices to be transmitted over a single channel. In 2000, there were three major digital transmission methods in use in the U.S.:

- Code Division Multiple Access (CDMA)
- Time Division Multiple Access (TDMA)
- Global System for Mobile Communications (GSM)
- Also IDEN, a proprietary system used by Nextel/Motorola (similar to TDMA and GSM but not interoperable). Nextel had 6.2 million subscribers as of 3Q 2000.

CDMA

CDMA is a spread-spectrum approach to digital transmission. This method of transmission assigns unique codes to each transmission, and then transmits over the entire spectrum. The mobile phone is instructed to decipher only a particular code in order to receive the designated transmission. The assignment of separate codes allows multiple users to share the same air space. CDMA increases capacity 8–20 times versus analog cellular, and also has greater capacity than TDMA. Figure 2.3 depicts how CDMA separates transmissions.

Since CDMA was not ready for deployment until after TDMA, it was first adopted by many of the PCS license winners (PCS licenses were distributed after cellular licenses). The U.S. and Korea are the largest CDMA markets worldwide. The primary

CDMA carriers in the U.S. are Sprint PCS and Verizon Wireless (Vodafone/Airtouch/Bell Atlantic/GTE). CDMA's major supporter has been Qualcomm, which holds a number of key CDMA patents. Figure 2.4 shows CDMA subscribers in the U.S.

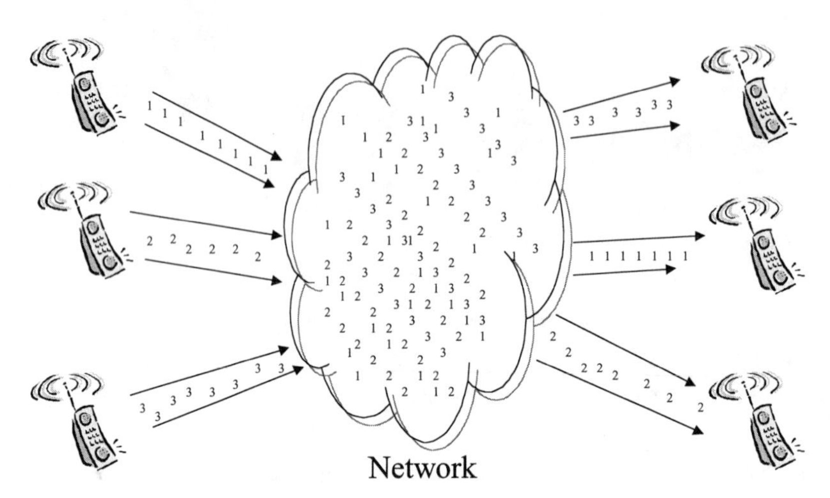

Figure 2.3
CDMA Operating Schematic.

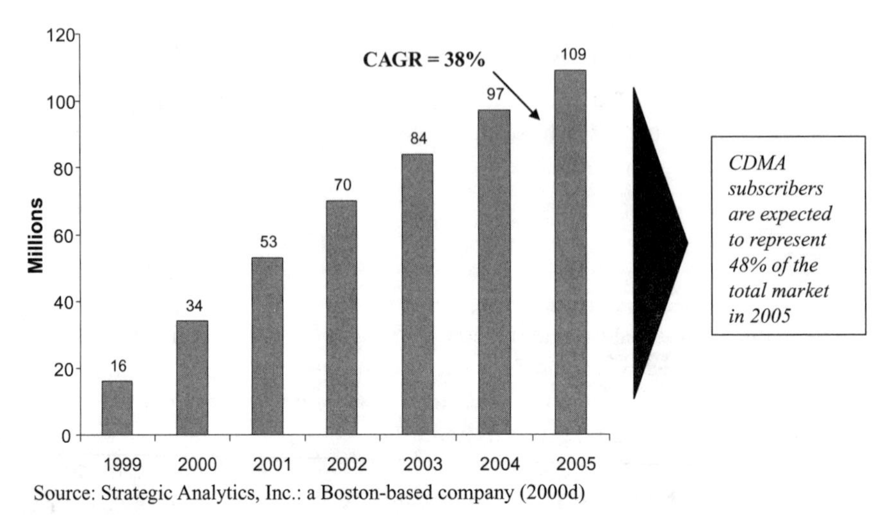

Figure 2.4
U.S. CDMA Subscribers.

CDMA is expected to grow rapidly in the U.S. due to its capacity advantages. It is also expected to be the basis for the two most likely 3G technologies, W-CDMA and CDMA 2000.

TDMA

TDMA uses a scheme in which the transmission channel is broken into six time slots (see Figure 2.5). Three of the time slots are used to carry information and three of the time slots are unused to minimize interference or noise. Work is currently underway to increase the number of time slots and thus capacity. TDMA increases capacity three to five times versus analog cellular.

The major TDMA carriers in the U.S. are AT&T Wireless and SBC/Bell South, now called Cingular Wireless. Figure 2.6 shows TDMA use in the U.S.

GSM

GSM is a similar transmission method to TDMA, in that it assigns time slots to communications. This is also true of the IDEN system employed by Nextel and supplied by Motorola. GSM is the standard adopted by Europe and much of the rest of the world, but in 2000 was the least widely deployed of the three major technologies in

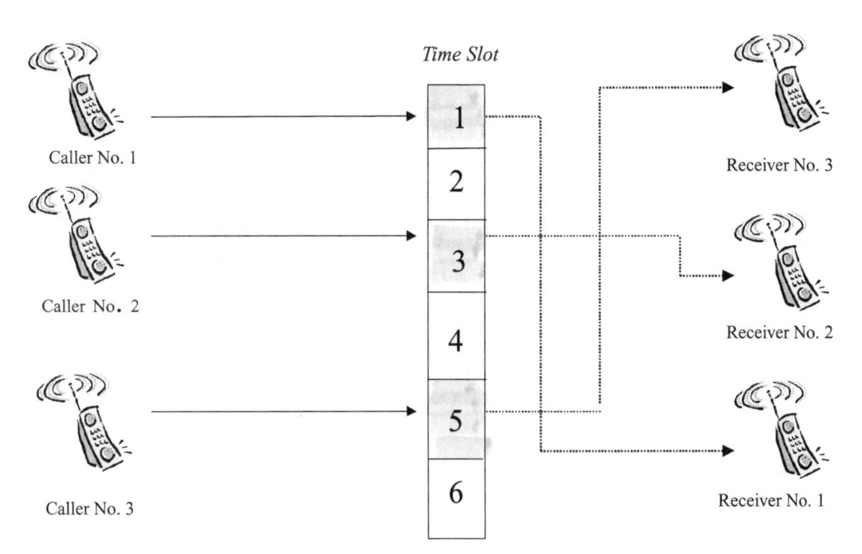

Figure 2.5
TDMA Operating Schematic.

the U.S. GSM is a complete standard, as it governs features in addition to the air interface, such as SMS messaging, and enables certain location technologies such as Enhanced Observed Time Difference and Enhanced Cell ID. Figure 2.7 shows GSM in the U.S.

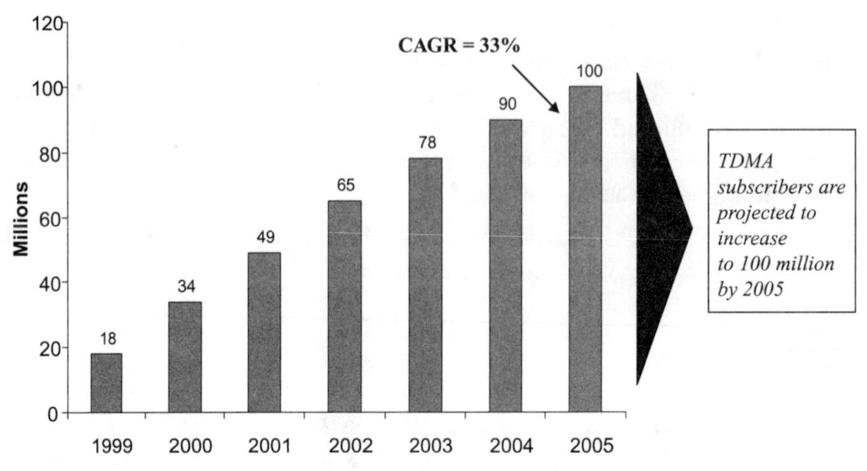

Source: Strategy Analytics, Inc.: a Boston-based company (2000d)

Figure 2.6
U.S. TDMA Subscribers.

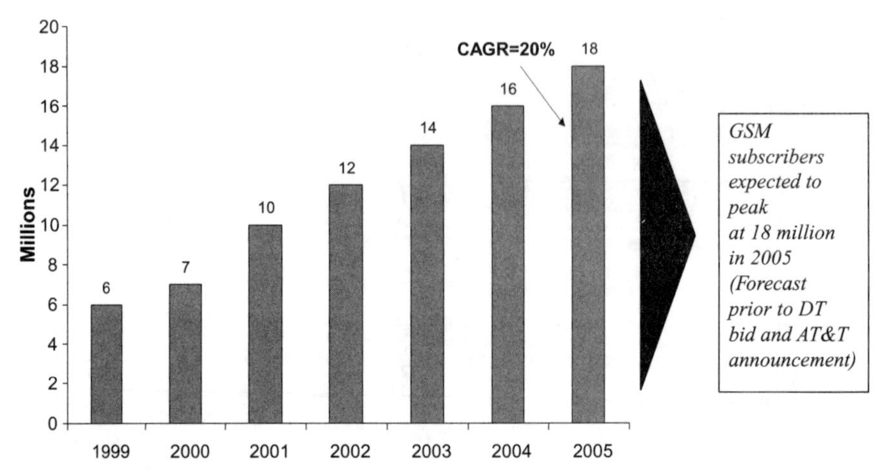

Source: Strategy Analytics, Inc.: a Boston-based company (2000d)

Figure 2.7
U.S. GSM Subscribers.

The major GSM carrier in the U.S. is VoiceStream. VoiceStream built a national presence by acquiring many of the smaller GSM carriers. In 2000, Deutsche Telecom (DT) bid to acquire VoiceStream, subject to government approval. As a European carrier, DT has a GSM network in Europe that it could then extend to the U.S., potentially creating global synergies.

On November 30, 2000, perhaps in response to this threat from DT, AT&T Wireless announced they would be building an entirely new GSM/GPRS/EDGE/W-CDMA network, alongside their TDMA network across a large portion of their coverage area, to provide a better migration path to 3G. They plan to cover 40% of their network with GSM/GPRS by the end of 2001, and 100% by the end of 2002. They have also announced that they will begin to deploy EDGE by the end of 2001, and W-CDMA, potentially in the 2003–2004 time frame.

CIRCUIT-SWITCHED TO PACKET-SWITCHED CONVERSION

All 2G technologies are circuit-switched systems, requiring dialing and establishing a dedicated circuit between the caller and the receiver. This dedicated circuit is maintained for the duration of the call, regardless of the actual quantity of information transmitted. This is similar to the current landline or Plain Old Telephone System (POTS) phone system.

A packet-switched system transmits data in packets that are reassembled by the receiver, rather than by establishing a dedicated connection. This is the same method used by the Internet. Packet-switching allows for an "always on" connection and a vast increase in capacity, since a dedicated circuit contains much more bandwidth than is consumed by a voice call. Packet-switching is a critical enabler for most of the data services currently being contemplated, and it allows for the "push" of information to the customer. Packet-switched systems such as GPRS were being piloted in Europe in late 2000.

General Packet Radio Service

THE EVOLVING WIRELESS TRANSMISSION TECHNOLOGIES

The following are terms likely to be encountered when referring to the evolution of wireless transmission technologies, and a brief explanation as to their meaning.

HSCSD

High Speed Circuit Switched Data (HSCSD) offers speeds of up to 58 kbps through a software upgrade to GSM networks. HSCSD is not a packet-switched technology, but works by combining transmission channels to improve data transmission. HSCSD's major disadvantage is a reduction in network voice capacity. The limitations of HSCSD make it an unlikely evolution path given the more robust capabilities of GPRS. No U.S. carrier has announced that it will be implementing this solution.

CDPD

Cellular Digital Packet Data (CDPD) is a packet-switched data transmission technology developed for use on cellular frequencies and is considered a precursor to GPRS. CDPD is a cellular network overlay that uses unused channels (in the 800-MHz to 900-MHz range) to transmit data in packets. CDPD offers data transfer rates up to 19.2 kbps.

AT&T's PocketNet service and the Novatel and Sierra PDA modems use CDPD. Other carriers supporting CDPD include Verizon, ALLTEL, and Ameritech. CIBC estimates that there were 600,000 individuals using CDPD services in the third quarter of 2000.

GPRS

General Packet Radio Service (GPRS) is a 2.5G technology that allows networks to send "packets of data" at rates up to 115 kbps. GPRS allows "always on" connections to send information immediately to the subscriber, with no dialup required. GPRS is more efficient than sending data over a circuit-switched wireless connection and will allow users to be charged per packet of data rather than by connection time.

GPRS is a data-only packet network overlay for GSM networks and is a relatively straightforward upgrade to existing networks, requiring a software and chip board upgrade. GPRS also requires a GGSN (Gateway GPRS Support Node), which is a packet router, and an SGSN (Serving GSN Support Node), which tracks the subscriber and provides security.

In 2000, the only handset available for GPRS services was from Motorola, and only in limited quantities. Although 115 kbps is the advertised speed for GPRS, handsets have not yet been able to achieve these speeds. Apparently, heat generation has been a serious problem, even at much slower speeds. GPRS handsets are expected to be more widely available in the second half of 2001, when Nokia is scheduled to release its product.

AT&T Wireless has announced it will begin to roll out a new GSM/GPRS network in 2001.

EDGE

Enhanced Data Rates for GSM Evolution (EDGE) is an evolutionary path to 3G services for GSM and TDMA operators. It represents a merger of GSM and TDMA standards and builds on the GPRS air interface and networks.

EDGE is a data-only upgrade and supports packet data at speeds up to 384 kbps. EDGE is able to achieve increased data transmission speeds through a change in its modulation scheme, from GMSK to 8 PSK. The upgrade to EDGE is relatively expensive and requires carriers to replace the transceivers (radio antennas) in every cell site. According to Commonwealth Associates this can cost as much as 60% of the original network cost (Greiper and Ellingsworth 2000, 21).

AT&T Wireless has indicated that they plan to migrate to EDGE.

3G TECHNOLOGIES

It appears that there will be at least two separate 3G solutions for the U.S. market (see Figure 2.8), preceded by several intermediate steps. Without a national standard, the U.S. will continue to experience increased costs and interoperability problems.

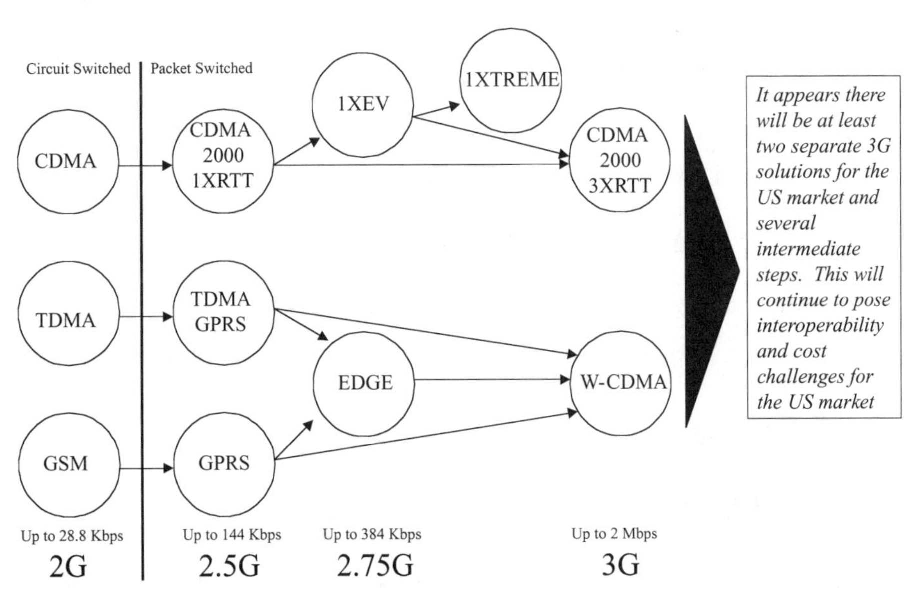

Figure 2.8
Technology Migration Paths to 3G.

3G-UMTS (also known as IMT-2000)

Universal Mobile Telecommunications System (UMTS) is a 3G technology standard (broadband, packet-switched) being developed as a global interoperable system for mobile wireless communications (similar to GSM in Europe). As of 2000, 200 wireless industry members had formed the UMTS Forum to discuss key issues and standards regarding UMTS. UMTS will continue to increase speeds of data transfer (up to 2 Mbps) and allow for additional value-added services. It is important to note that the terms 3G, UMTS, and IMT-2000 are often used interchangeably.

In industry parlance, 3G generally refers to packet-switched transmission rates up to 144 kbps in a mobile vehicular environment, 384 kbps in a mobile pedestrian environment, and up to 2Mbps in a stationary environment. 3G should also provide multimedia capabilities, deliver wire-line voice quality and security, support simultaneous voice and data connections, and provide location information.

European governments have auctioned UMTS licenses to service providers. These auctions brought in billions of dollars more than expected in both Germany and the UK, showing the high expectations carriers have for the success of broadband wireless services.

Great Britain's auction brought in $35.5 billion for five licenses. Vodafone/AirTouch had the highest bid, paying more than $9 billion for 35 MHz of spectrum. Six German licenses of 20 MHz each were sold for a total of over $45 billion. In addition to the cost of the spectrum, there are very heavy capital expenditures required to begin service. This makes 3G a very risky proposition for the European carriers because of the uncertainty surrounding key services and the amount people are willing to pay for them.

To make more spectrum available in the U.S., the FCC has scheduled two auctions (see Chapter 1). The first is the reauction of 1.9-GHz spectrum, that began on December 12, 2000. This spectrum is particularly valuable since it can be used almost immediately after winning the auction. The second is the auction of 700-MHz spectrum scheduled to begin September 12, 2001. There are many issues surrounding this spectrum and this auction has a high potential of being delayed again.

Although the goal of UMTS is to have one standard, several are currently being considered. If several standards are adopted, interoperability will be the key issue. The two major 3G standards appear to be CDMA2000 and W-CDMA (Wideband CDMA).

CDMA2000

CDMA2000 is an evolution to 3G for CDMA networks. CDMA2000 is compatible with current CDMA networks, IS-95A and IS-95B (data-enabled CDMA).

CDMA2000 is expected to be rolled out in two phases, 1X or 1XRTT and 3X or 3XRTT (the abbreviation RTT stands for Radio Transmission Technology). 1XRTT is expected to provide a packet data rate of 144 kbps in a mobile environment, and operate on 1.25 MHz of bandwidth (1X). It is also forecast to lead to a two-fold increase in voice capacity and battery standby time. Commercial availability is expected in 2001.

1XRTT is a relatively straightforward upgrade requiring a software and chip board upgrade to the network. According to Yankee Group this is likely to cost about 30% of the original network cost, which is substantially less than the cost of upgrading other networks and has the advantage of substantially increasing voice capacity (Greiper and Ellingsworth 2000, 17). Sprint PCS and Verizon have announced they will use 1XRTT. Sprint PCS plans to roll out 1XRTT in late 2001 and expects the cost to be $700–1,000 million.

Additional intermediate CDMA solutions have also emerged. These include 1XEV, also known as HDR (High Data Rate), and 1Xtreme. 1XEV, from Qualcomm, is estimated to support transmission speeds in excess of 2 Mbps. 1Xtreme is a proposal from Motorola and Nokia that is reputed to support peak speeds over 5 Mbps and steady state speeds of 1–2 Mbps.

3XRTT promises data-rate increases to 2 Mbps and will use 3.75 MHz of bandwidth (3X). Commercial availability for 3XRTT is expected in 2002. However, with the emergence of higher speed intermediate solutions, the need to move to 3XRTT may be reduced. As of January 1, 2001, no carrier had committed to 3XRTT.

W-CDMA

Wideband Code Division Multiple Access (W-CDMA) is supported by GSM operators as their 3G technology, and is viewed as the best evolutionary path for GSM/TDMA operators. W-CDMA will support speeds of 384 kbps–2 Mbps and will use a faster chip than CDMA2000. W-CDMA will not be as compatible with early versions of CDMA as CDMA2000, since the W-CDMA chip is set up for the timing of GSM rather than the timing of CDMA.

W-CDMA uses a wide channel, 5 MHz, and requires a complete network overhaul. The installation cost for W-CDMA in estimates cited by Commonwealth Associates is 100–120% of the original network cost (Greiper and Ellingsworth 2000, 22).

NTT DoCoMo, Japan's largest wireless carrier, is expected to be the first carrier to implement W-CDMA. DoCoMo plans to deploy W-CDMA in select metropolitan markets in the spring of 2001. This launch was subsequently delayed until the fall of 2001 due to issues surrounding call handoffs between cells. AT&T Wireless has committed to W-CDMA, with probable deployment in the 2003–2004 time frame, to enable international roaming with their partner NTT DoCoMo. NTT DoCoMo purchased 16% of AT&T Wireless in November 2000.

Outlook

There is tremendous promise in the new broadband packet-switched network transmission technologies. At the advertised speeds, they will be substantially faster than many cable modem and DSL speeds in 2000 and can enable a broad range of mobile multimedia services such as streaming video and audio. These speeds would also enable large file transfer and a number of other key corporate applications.

The promise of these technologies has caused great excitement among marketers, software developers, and the entire wireless industry. However, a word of caution is in order. Current network speeds are relatively slow and coverage can be spotty, as anyone who uses a wireless phone frequently can attest. Additionally, advertised data speeds are generally much higher than one can reasonably expect to experience in real-world use (see Figure 2.9 for various uses and the required transmission speeds). This is due to such issues as distance from the base station, number of other users on the system, and error correction.

There is also substantial risk that the deployment of these types of broadband networks will be delayed until the business model has proven itself. The cost of spectrum and the level of capital expenditure are immense. Additionally, there is not a clear consensus in the industry as to what the "killer app" will be, and how much people

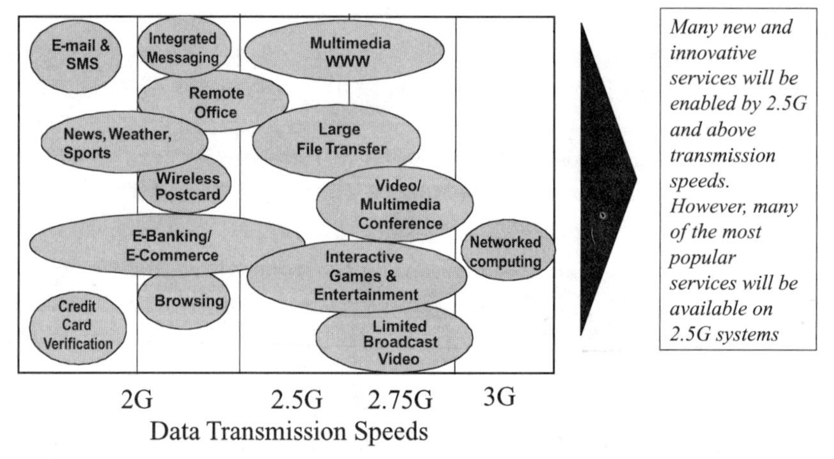

Figure 2.9
Data Transmission Speed Requirements.

will be willing to pay for these services. The U.S. has an advantage in this regard, as the Japanese and Europeans will likely introduce their services first, providing clearer evidence as to the potential for different applications.

Handsets are also likely to be a limiting factor in achieving 2.5G and 3G services. GPRS handsets have not come close to achieving advertised data transmission speeds, due to heat-generation issues. Handsets are also likely to be an important issue in realizing 3G data speeds and applications. In fact, NTT DoCoMo has already expressed concern over the handset issue for their 2001 launch of W-CDMA in Japan.

W-CDMA is likely to be the most popular 3G standard worldwide, and will gain a significant presence in the U.S. through AT&T Wireless, Cingular, and VoiceStream/Deutsche Telecom. However, in the U.S., there is also a substantial CDMA camp that will adopt CDMA 2000. These carriers include Verizon Wireless and Sprint PCS. It appears that the evolution path for the CDMA operators is relatively easier and less expensive.

In assessing the demand for broadband wireless services and people's willingness to pay, it is helpful to remember the plight of Iridium, and the promise of video on demand from the cable industry in the early 1990s. In both cases, the price customers were willing to pay was much less than the amount needed to justify the investment.

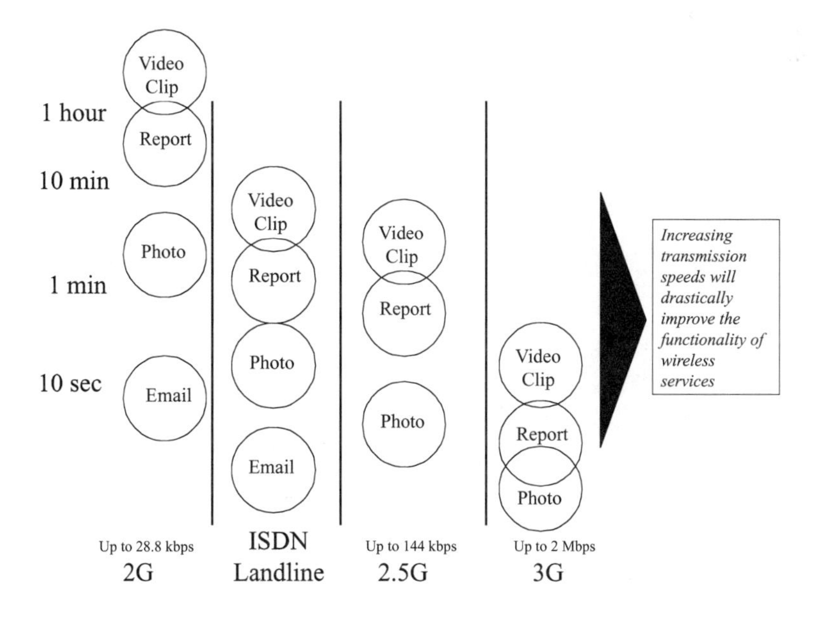

Source: UMTS, 2000

Figure 2.10
Transmission Times over Mobile Networks.

In the U.S., broadband wireless services will be deployed, albeit on a more conservative timetable than the most enthusiastic would have one believe. Broadband, packet-switched wireless services are likely to be available in the U.S. sometime after 2003. The most likely scenario is that these services will be adopted first by businesses in the major metropolitan areas to improve productivity and customer service, and then migrate to consumers for entertainment and communication. Once in full deployment, these services will offer enormous benefits to society and tremendous increases in personal capability.

PAGING/DATA NETWORK TECHNOLOGIES . . . . . .

Paging/data networks are mobile wireless communications networks that communicate data rather than voice information to the subscriber. They operate under the same principles as voice networks, where the radio signal is transmitted from a base station to a mobile terminal. The major data networks today are Mobitex, DataTac (ARDIS), and the various paging networks throughout the U.S. The vast majority of the U.S. paging networks operate on the FLEX protocol, or its upgrade ReFLEX.

CDPD is another data network. However, it is more of an overlay of the cellular network than a stand-alone data network. Additionally, 2.5G and 3G networks can be thought of as data networks (discussed in the previous section). Ricochet, by Metricom, is another wireless data network, but it did not support mobility in 2000, and thus, is not addressed in this text.

Some of the major advantages of paging/data networks are extensive nationwide coverage, superior in-building reception, ease of use, smaller terminals, and longer battery life.

Mobitex

Mobitex is a digital, two-way packet-switched network protocol that operates on cellular frequencies to deliver packetized data. This standard was developed by Ericsson and Swedish Telecom Radio in the early 1980s. The Mobitex standard was subsequently opened up to other manufacturers in 1994. Nortel, Lucent, and Hughes Network Systems have entered the market and supply equipment based on this protocol.

BellSouth Wireless Data (BSWD) is the dominant Mobitex firm in the U.S., and their service covers 93% of the U.S. urban population (see Figure 2.11). BSWD offers proprietary service for Palm.net, the integrated wireless service for the Palm VII, and also provides service for the RIM BlackBerry two-way pager. The BellSouth network operates in the 896–901 MHz and 935–940 MHz spectrum bands. According to Gartner Inc., there were about 300,000 Bell South Wireless Data subscribers in 2000 (Redman 2000, Fig. 4.45).

Mobitex is a secure system with error correction and the ability to request retransmission of lost packets. The service can be billed by the carrier on a per-packet basis instead of being based on connection time. This is a large advantage for data services. Since Mobitex is packet based, the service is always on, and can push information to the subscriber. Mobitex supports transmission rates of 8 to 9.6 kbps and up to 19.2 kbps with newer base station hardware and modems. Mobitex allows for nationwide and international roaming.

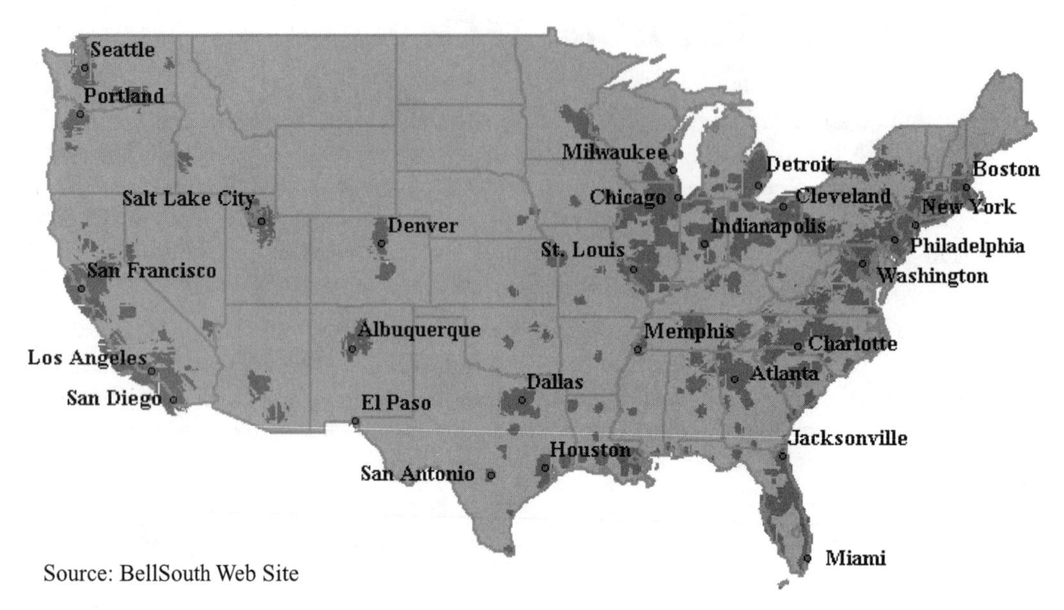

Figure 2.11
BellSouth Wireless Data Coverage Map.

DataTac

DataTac was developed by Motorola and IBM to give IBM service employees remote access to the company mainframe. The service, named ARDIS, commenced in 1990. ARDIS is very similar to Mobitex, in that it is a digital, two-way packet-switched data network. ARDIS supports data speeds up to 19.2 kbps and covers 80% of the U.S. population, including Puerto Rico and the Virgin Islands.

Since the service was designed to support IBM service personnel, a primary goal was to provide superior in-building service. This is one of the key advantages of this network versus cellular networks.

The ARDIS network was acquired by Motient on March 31, 1998. However, the network is still predominantly used by IBM employees. The ARDIS network also supports the BlackBerry pager (850) and roaming between Mobitex and ARDIS has been enabled. According to Gartner Inc., there were about 150,000 Motient subscribers in 2000 (Redman 2000, Fig. 4.45).

Paging

Paging was one of the first mobile wireless communications applications and gained in popularity quickly. It was first adopted by doctors and business executives. As prices fell, it spread to the masses. Motorola has been the dominant firm in the paging industry, both in terms of infrastructure and terminals (pagers) for many years. However, in 2000, Motorola exited the manufacture of paging infrastructure and licensed the ability to make its equipment to Glenayre Technologies.

Paging adopted a common paging code very early on. This code was called Post Office Code Standardization Advisory Group (POCSAG). This standard was then superseded by the FLEX standard from Motorola (discussed in more detail below). FLEX offered increased speed and throughput, greater capacity, and higher reliability. The FLEX standard has captured more than 90% of the worldwide paging market.

The capabilities of paging have evolved and improved rapidly. Paging service has advanced from tone to numeric to alpha-numeric and now to full two-way. The paging market is segmented by the level of communication capability, with interactive paging growing at the expense of one-way paging. The classes of service can be defined as one way, 1.5 way (where the device sends a receipt back over network), 1.7 way (where the device has two-way communication with an optional preset response), and full two way (where original messages and responses can be transmitted using a QWERTY keyboard).

According to Gartner, the top seven paging companies have more than 33 million subscribers and IDC estimated that there were a total of 41.5 million paging subscribers in 1999 (Redman 2000, Figs. 4.45, 4.44). However, the total number of subscribers declined in 2000, despite the growth of two-way paging.

FLEX

FLEX is a wireless technology developed by Motorola and introduced in 1993. FLEX improves channel efficiency, reduces the cost of paging networks, and enables new services. FLEX supports data speeds of up to 6.4 kbps. FLEX and its derivatives command over 90% market share. FLEX and ReFLEX are not packet-based systems.

ReFLEX

ReFLEX is a follow-on to FLEX that enables two-way alpha-numeric paging through a response channel. ReFLEX enables an immediate response from the terminal and guarantees message delivery.

There are two versions of ReFLEX available, ReFLEX 25 and ReFLEX 50. ReFLEX 25 receives data at 9.6 kbps and transmits at up to 6.4 kbps. ReFLEX 50 receives data at 25.6 kbps and transmits at up to 9.6 kbps.

The @ctiveLink module for the Handspring Visor works on Metrocall's ReFLEX 25 network. This allows Handspring users to send/receive messages, access the Internet, and send/receive email.

Outlook

With the Mobitex and ARDIS networks and the advent of ReFLEX (see the comparison in Table 2.1), it seems clear that the future of paging/data networks is in two-way communication versus traditional one-way numeric paging. These services face very stiff competition from the mobile wireless telephony carriers, but due to some of the inherent advantages of paging/data networks—extensive coverage, in-building penetration, and extended battery life—they are likely to continue to remain viable, particularly if they expand their applications to include telemetry and monitoring.

Telemetry is machine-to-machine or machine-to-human communication. The surface has barely been scratched in this market and paging's capabilities make a good fit. For instance, monitoring in-building machines, like copiers or vending machines, could be valuable. Monitoring applications require two-way communication, but generally need little bandwidth.

Due to the challenging economics of the industry there are not likely to be any major new paging/data networks erected for the foreseeable future (not including 2.5G and 3G telephony networks). It is also probable that there will be further consolidation in the paging service provider market.

Table 2.1 ReFLEX versus Mobitex.

	ReFLEX	Mobitex
Architecture	2-way network overlay	Integrated single design
Latency	Delay	Real time
Capacity	Lower bandwidth/ small messages	Higher bandwidth/ larger messages
Coverage	Excellent	Good
Device Price	$200-350	~$500
Efficiency	Broadcast capability	Message by message
Development	Open, good support	More systems integration
User Pricing	Cheaper	More expensive
Infrastructure cost	Cheaper	More expensive

Source: Strategy Analytics, Inc.: a Boston-based company (1999)

3 Network Suppliers

In this chapter...

After spectrum has been acquired and a transmission technology has been selected, the network must be built. There are two primary components of the mobile wireless communications system—terminals and infrastructure. The wireless carrier must select the network infrastructure to deploy and the terminals that will be allowed to operate on their network. Firms that supply terminals and infrastructure can be classified as Network Suppliers. These firms hope to profit by supplying equipment or products to the wireless network.

Terminals are devices that send and receive radio signals between the subscriber and the wireless network. The primary classes of mobile wireless communication terminals are wireless handsets or telephones, pagers, and Personal Digital Assistants (PDAs).

Mobile wireless communications infrastructure is the equipment that enables wireless communication between mobile terminals and other terminals, either mobile or landline. This infrastructure consists of several key elements: the switching system, the base station system, and the operation and support system.

Also included in the infrastructure category are location technologies since these are either network-based, handset-based, or hybrid (combination of network and handset) capabilities that are enabled by network suppliers.

The network supplier market is very large, although much smaller than the carrier market. The different segments of the market are unique and face substantially different market conditions and attractiveness, despite substantial company overlap in market participation and customers. For example, Ericsson is a leading firm that supplies both infrastructure and handsets. In 2001, Ericsson's infrastructure business was very strong and had great success in winning contracts in 2.5G and 3G networks; however, they were forced to substantially retreat from the handset business and outsource all of their handset manufacturing.

Based on the assessment of the structural attractiveness of the two markets, this is not a surprising outcome. Infrastructure is one of the more structurally attractive markets in the mobile wireless communications industry while handsets are much less structurally attractive. Where terminals are concerned, PDAs appear to be the most structurally attractive market.

TERMINALS

Mobile wireless terminals are a huge and rapidly growing market in the U.S., and are expected to exceed 180 million units per year by 2005 (see Figure 3.1). The primary classes of terminal are wireless handsets, wireless PDAs, and pagers. Wireless handsets are by far the largest category, followed by pagers. However, the most structurally attractive terminal market appears to be the PDA market.

When looking at longer term forecasts for the total mobile wireless communication device market, it may be helpful to look at the total requirement for wireless access devices, with the ultimate form factor being determined by the utility to the customer. Thus, clever manufacturers may be able to influence the relative results of a particular area by offering a superior product. For instance, wireless PDAs are not forecast to make up a large portion of the overall market; however, a very successful product could double the size of the wireless PDA market without dislocating much of the total market. Additionally, wireless handsets can be expected to incorporate much of the PDA functionality.

Ultimately, the wireless access terminal is likely to evolve into a digital control device through its access to data, the Internet, location information, and the incorporation of Bluetooth or similar technology.

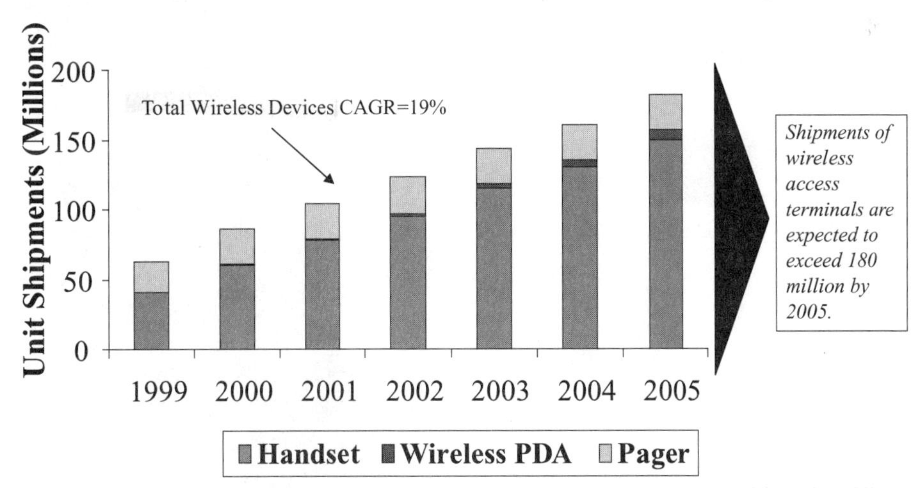

Source: Adapted from IDC and Strategy Analytics, Inc.: a Boston-based company (2000d, 1999, 1998)

Figure 3.1
Unit Shipments of U.S. Mobile Wireless Terminals.

MOBILE WIRELESS HANDSETS . . . .

A wireless handset, commonly referred to as a cell phone, is a device used to receive and transmit voice—and now data—information in a mobile wireless environment (see Figure 3.2). Data communication is expected to become a much more important functionality going forward. However, the true "killer app" for the wireless handset is voice communications.

Figure 3.2
Some Selected Mobile Wireless handsets. Reproduced with permission from Motorola, Inc. © 2001, Motorola, Inc.

Market Size and Projections

The mobile wireless handset market is growing extremely rapidly, driven by strong subscriber gains, the transition from analog to digital, a high level of industry churn, and a rapid replacement cycle. There are currently more than 100 million wireless subscribers in the U.S. market, and each one requires a handset.

In 2000, the U.S. wireless handset market is expected to be in excess of 60 million units and more than $6 billion in revenues (see Figures 3.3 and 3.4). This represents 39% growth in units and 29% growth in revenues over 1999. The compound annual growth rate in units for the handset market between 1999 and 2005 is projected to be 24%, leading to an industry with revenues greater than $10 billion. This is a phenomenal growth rate for an industry of this size.

As the market matures, one can expect the bulk of handset sales to be driven by replacement purchases, particularly as data become a larger portion of wireless traffic. Improved design, reduced size, and improvements in battery life are also key factors in driving replacement sales.

According to a Strategy Analytics survey, in 2000, one-third of subscribers replaced their existing handset (Strategic Analytics 2000c). The main reasons cited for replacing phones were:

Upgrade to digital	31%
Wanted smaller/lighter/latest	21%
Lost/broken/stolen	26%

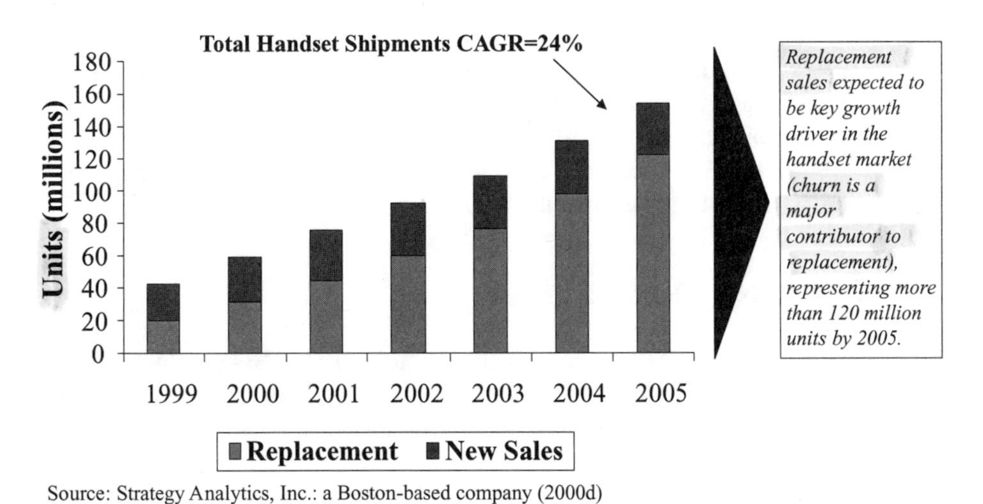

Source: Strategy Analytics, Inc.: a Boston-based company (2000d)

Figure 3.3
U.S. Wireless Handset Shipments.

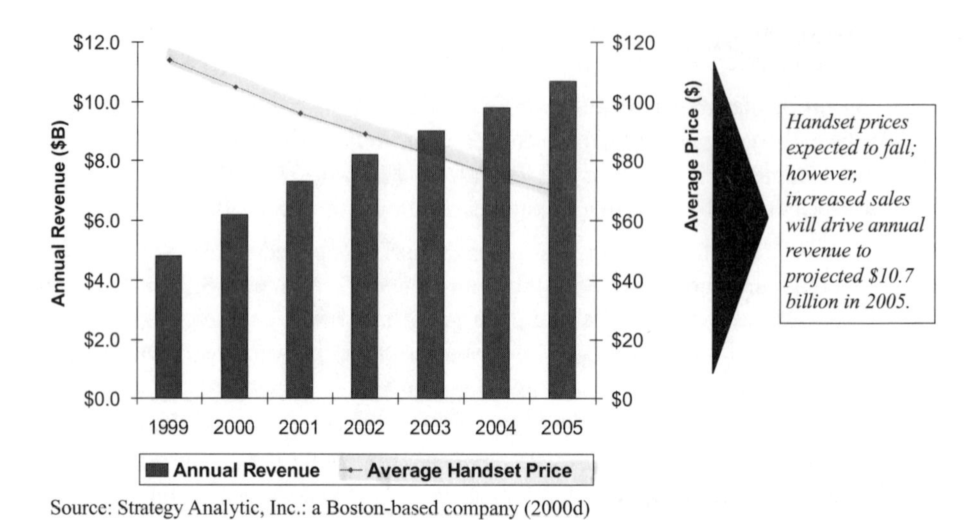

Source: Strategy Analytic, Inc.: a Boston-based company (2000d)

Figure 3.4
U.S. Wireless Handset Annual Revenue and Average Handset Price.

While industry growth is extremely strong, like most consumer electronics products the price of handsets is declining rapidly (see Figure 3.4). This is expected to continue in the future as competition intensifies and as more functions are consolidated onto a single chip.

The handset manufacturing industry is characterized by intense competition, and the level of competition is expected to continue to intensify as more companies enter the market. Many companies feel that wireless handsets will become a key computing, multimedia, and control device. Consequently, they are very anxious to capture a large share of this rapidly growing pie and may be willing to sacrifice profitability for market share. Some of the likely entrants into this market would be consumer electronics companies such as Philips, Siemens, and Sony (all manufacture phones, but not for the U.S. market in 2000); the PDA manufacturers such as Palm and Handspring; and such computer manufacturers such as Compaq and Hewlett Packard.

Handset Market Share

The Big 3 global handset manufacturers currently are Nokia, Motorola, and Ericsson. This is also generally true for the U.S. market, with the exception of CDMA technology, where Qualcomm (now Kyocera), Audiovox, and Samsung have been able to capture significant share.

Qualcomm is the primary proponent of CDMA and holds many key patents on this technology. It sold its handset manufacturing business to Kyocera in early 2000.

It is also not surprising to see Samsung, a Korean manufacturer, among the top CDMA phone suppliers, since Korea is one of the largest CDMA markets in the world.

The market share charts in Figures 3.5–3.9 depict the intense level of competition with six relatively strong competitors in the U.S. market and more on the way. Nokia clearly stands out as the winner in 2000, and it is reflected in their industry-leading operating margins.

Power of the Carrier

The dominant feature of the U.S. handset market is the tremendous power of the carrier. The carrier is both the primary customer and the primary channel to the consumer. The carrier also provides a large subsidy to the retail price of the handset, on the order of $50–$100 per unit in 2000, both in their own stores and through other retailers (like Circuit City and Radio Shack). This subsidy serves to stimulate increased demand for phones since consumers rightly feel that they are getting a $300 phone for $99 (subsidy at wholesale price has a larger impact on retail price).

One of the detrimental side effects of this subsidization, from a carrier perspective, is the high level of churn it creates. When customers want to get a new phone, they have only to switch carriers (although, at this time, their phone number will be

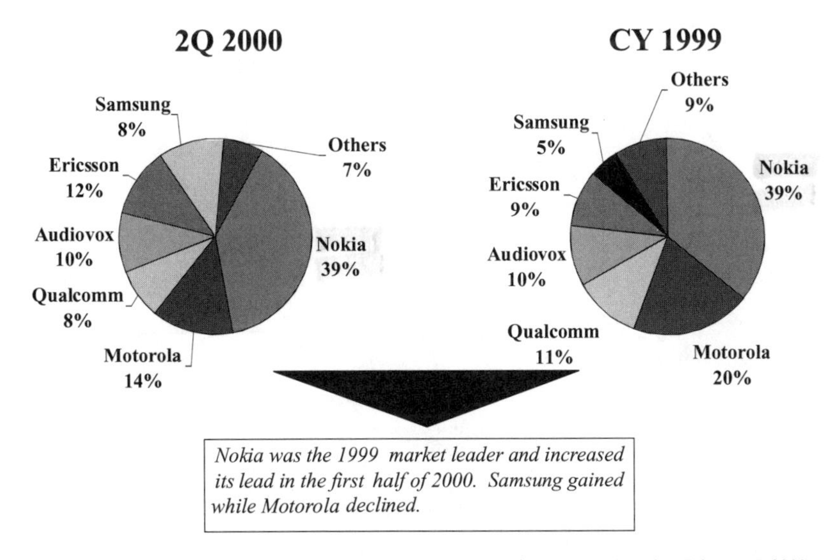

Statistical data from The North American Wireless Marketplace 2nd Quarter 2000 Update February 5, 2001, by Paul Dittner and Bryan Prohm, Gartner

Figure 3.5
U.S. Market Share for Sales of Mobile Handsets to End Users.

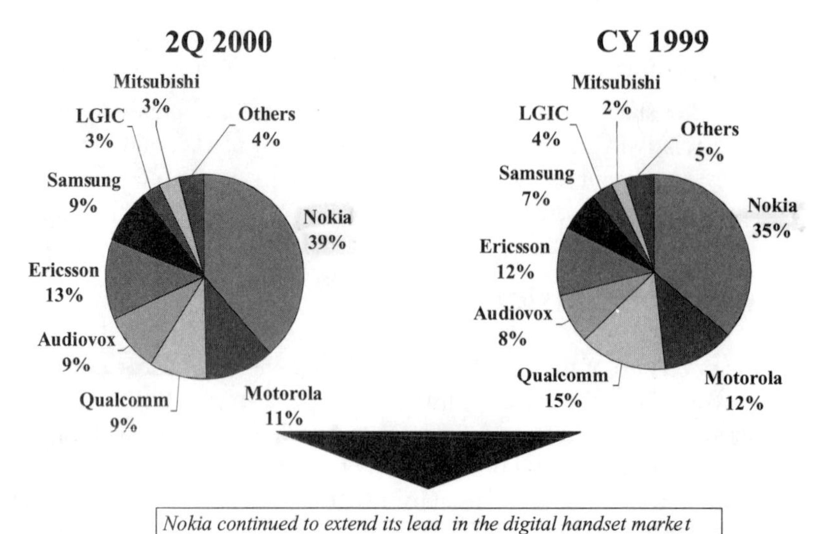

Statistical data from The North American Wireless Marketplace 2nd Quarter 2000 Update February 5, 2001, by Paul Dittner and Bryan Prohm, Gartner

Figure 3.6
U.S. Market Share for Sales of Digital Handsets to End Users.

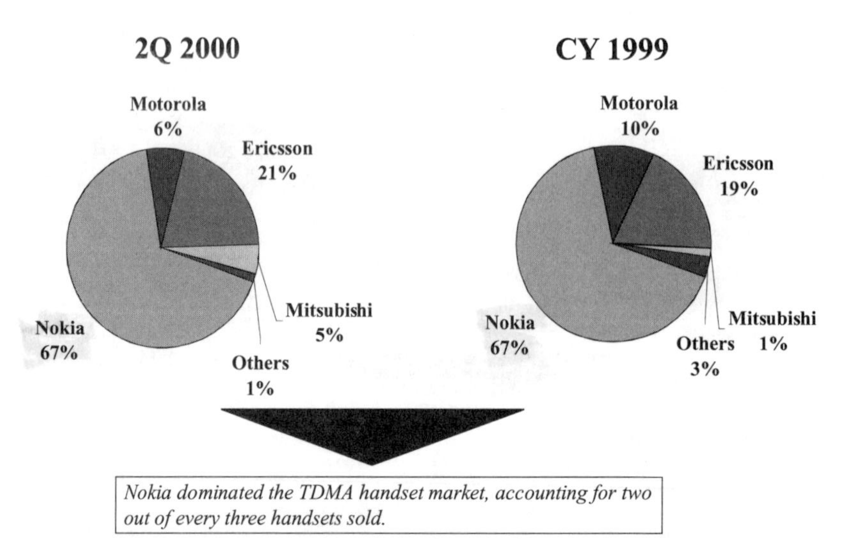

Statistical data from The North American Wireless Marketplace 2nd Quarter 2000 Update February 5, 2001, by Paul Dittner and Bryan Prohm, Gartner

Figure 3.7
U.S. Market Share for Sales of TDMA Handsets to End Users.

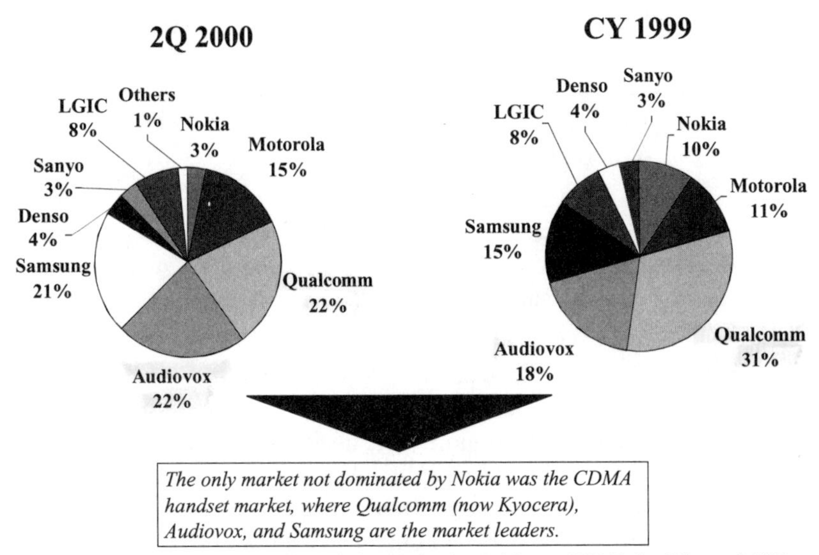

Statistical data from The North American Wireless Marketplace 2 nd Quarter 2000 Update February 5, 2001, by Paul Dittner and Bryan Prohm, Gartner

Figure 3.8
U.S. Market Share for Sales of CDMA Handsets to End Users.

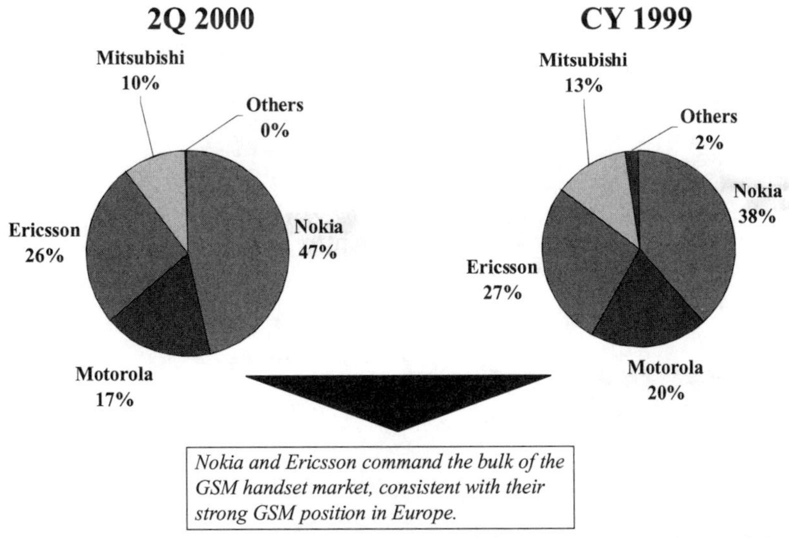

Statistical data from The North American Wireless Marketplace 2 nd Quarter 2000 Update February 5, 2001, by Paul Dittner and Bryan Prohm, Gartner

Figure 3.9
U.S. Market Share for Sales of GSM Handsets to End Users.

changed too), or threaten to switch and see if their incumbent carrier responds in kind. This is a large driver of replacement sales and a boon to handset manufacturers, because each time new service is activated a new handset is required by the carrier. In 2000 there was no ability to keep one's current handset when switching carriers. Churn is discussed more thoroughly in Chapter 4.

According to a 1999 IDC survey, 42% of wireless handsets were obtained directly from the carrier (see Figure 3.10). Strategy Analytics found that 66% of customers reported obtaining their handsets from a carrier store in a 2000 survey (see Figure 3.11). This is an extremely important dynamic in the marketplace and has a large impact on handset manufacturer profit margins.

Carriers are large, sophisticated buyers who influence very large volumes of handsets both through their qualification of the phone on their network and through which handset they decide to promote most heavily. A good example of this has been the promotion of the Nokia handset by AT&T for their Digital One-Rate Plan. Through this arrangement, Nokia was able to secure two-thirds of the U.S. TDMA market in 1999. In almost every commercial or promotional piece during AT&T's large marketing campaign, the Nokia phone was featured, making the Nokia virtually synonymous with the One Rate Plan.

It should be noted that the failure of Ericsson and Motorola to have strong products in the TDMA category in 1999 benefited Nokia as well.

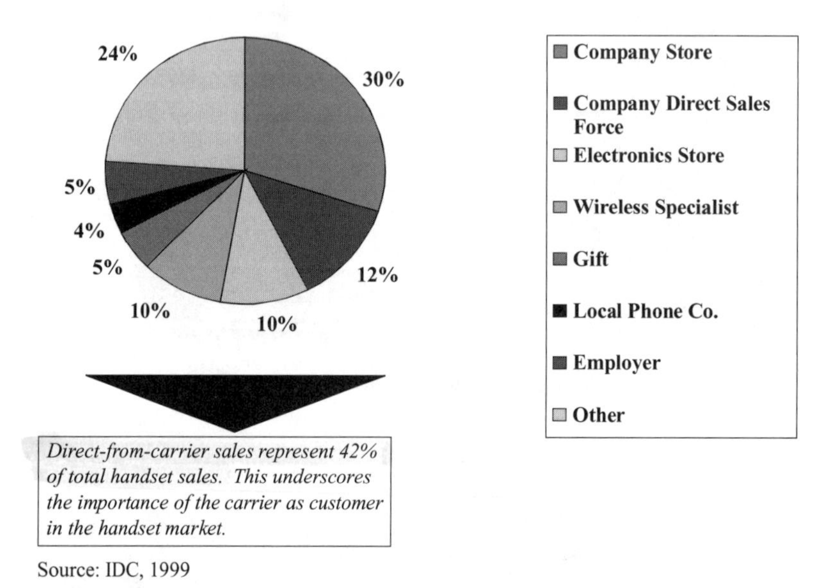

Source: IDC, 1999

Figure 3.10
Source of Most Recent Handset Purchase.

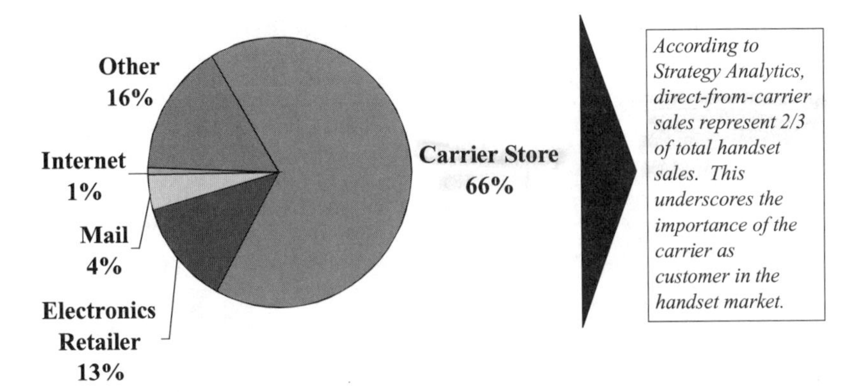

Source: Strategy Analytics, Inc.: a Boston-based company (2000c)

Figure 3.11
U.S. Handset Distribution.

Differing Transmission Technologies

The U.S. mobile wireless communications market is characterized by a variety of different transmission technologies. This situation requires the development of an almost-custom phone for each network operator. This is a key difference between the U.S. and the European markets. The variety of transmission technologies increases handset research and development and network-specific testing costs, while reducing volumes of a particular model for manufacturers.

The different wireless transmission technologies (CDMA, TDMA, GSM, AMPS) and different frequencies (Cellular 800 MHz and PCS 1900 MHz) in the U.S. market require mobile handsets capable of supporting them. For example, AT&T Wireless subscribers must use tri-mode handsets capable of supporting AMPS at 800 MHz, TDMA at 800 MHz, and TDMA at 1900 MHz to ensure national coverage on the AT&T network and to allow off-network roaming. Table 3.1 shows the major carrier/handset manufacturer relationships.

As users travel (roam) from one region to another, the wireless handset automatically adjusts to the technology and frequency in use in that particular region, so that calls can be placed and received. It is prudent to remember, however, that when roaming not all digital features will work all the time, such as voice mail notification. Therefore, it is often necessary to speculatively call in for voice mail while roaming.

Sprint PCS, which operates an all-digital network at 1900 MHz, has to include AMPS capability in their handsets to allow roaming in the large parts of the country without digital PCS coverage. This is an example of a dual-mode phone. Dual-mode phones support two technologies while dual-band phones support two frequencies.

Table 3.1 Major Carrier/Handset Manufacturer Relationships.

Carrier	Handset Manufacturer	Technology
AT&T Wireless	Ericsson, Mitsubishi, Motorola, Nokia	TDMA
SBC/BellSouth	Ericsson, Motorola, Nokia	TDMA
Sprint PCS	Denso, Kyocera, Motorola, Neopoint, Samsung, Sanyo	CDMA
Verizon	Audiovox, Kyocera, Motorola, Samsung	CDMA
VoiceStream	Ericsson, Motorola, Nokia	GSM

The inclusion of analog capability in most U.S. phones is the main reason handsets are still so large in this country, versus the rest of the world.

Handset Components

Over time, handsets have become smaller and more feature rich. Much of the cost and functionality of the wireless handset is driven by powerful semiconductors that enable many value-added features. Semiconductors routinely account for over 50% of the cost of the handset. Some of the leading chip suppliers to the handset industry are Texas Instruments, Qualcomm (CDMA), and Intel. See Tables 3.2 and 3.3 for the cost and components of leading handsets.

Table 3.2 Qualcomm-CDMA Handset Bill of Materials

Typical Bill of Materials	Cost	Suppliers
Baseband ASIC	$18	QCOM, LSI, INTC, Internally Developed by Handset OEMs
Power Mgmt	$ 1	QCOM, TXN, STM
Flash Memory	$ 3	INTC, AMD, ATML, STM
RF/IF Chipset	$ 8	QCOM, RFMD, CNXT, TQNT
Power Amp	$ 5	RFMD, CNXT, ANAD, TQNT
Other	$ 5	
Total	$40	

Source: Bernstein Research

Table 3.3 GSM Handset Bill of Materials

Typical Bill of Materials	Cost	Suppliers
Baseband ASIC	$14	Internally Developed by Handset OEMs, TXN, MOT, STM, Philips, Infineon, CNXT, ADI
CODEC	$ 3	TXN, STM, ADI, Philips, Infineon
Power Mgmt	$ 1	TXN, STM, ADI, Philips, Infineon
Flash Memory	$ 3	INTC, AMD, ATML, STM
Smart Card	$ 2	STM, Infineon
RF/IF Chipset	$ 7	Philips, Infineon, RFMD, CNXT, TXN, ADI
Power Amp	$ 5	RFMD, CNXT, ANAD, TONT
Other	$ 3	
Total	$38	

Source: Bernstein Research

Batteries

The battery is another key feature of the wireless handset. Nickel cadmium (NiCad) batteries were the most widely used in wireless handsets in 2000. The advantages of NiCad batteries are that they are inexpensive, easy to charge, and operate in a wide temperature range. The major drawbacks to NiCad are the memory effect in recharging (if not completely discharged, over time they will accept less of a recharge thus shortening usage time) and that they are made with a toxic metal.

Nickel metal hydride (NiMH) batteries are gaining in popularity. The major advantages for this type of battery are greater energy density (energy per volume) and capacity, nontoxicity, and no memory effect. The main disadvantages are higher cost and susceptiblity to damage from overcharging.

Lithium ion batteries are becoming the technology of choice for many high-end handsets. The advantages of Li-ion are high energy density and high specific energy (energy per weight), low self-discharge, and fast charge capability (ability to get a significant percent of recharge quickly). Li-ion batteries have roughly twice the talk and standby time of NiCad. The main disadvantage of Li-ion is that overcharging can present a safety risk if not properly regulated.

The future, however, may belong to the polymer lithium ion battery. When this battery becomes commercially available it will offer tremendous shape flexibility, such as the dimensions of tape. This will allow for unprecedented flexibility in handset design.

SIM Cards

Subscriber Information Modules—or SIM cards—are removable cards that enable mobile users to customize their handsets and access the services of other carriers outside their home region. SIM cards can allow international travelers to access services from other operators by inserting the SIM card into a mobile phone appropriate for that region. SIM cards can also contain personal information such as address books, phone numbers, and calendars. They can store account or credit card numbers securely and transfer them from phone to phone, thus enabling m-commerce.

SIM cards benefit consumers through the ability to upgrade or replace their phones easily. Handset manufacturers may benefit since more units may be sold as consumers purchase different phones for different occasions, such as a lightweight phone for weekends and a heavy duty phone for the business week. Service providers should appreciate the improved ability for consumers to upgrade to new phones without changing service providers, thus potentially reducing churn.

In 2000 SIM cards were unique to the GSM countries; however, CDMA SIM phones were expected to be in production by the end of 2000 or early 2001.

Bluetooth

Bluetooth is a royalty-free global wireless technology standard. In early 1998 a group of computer and telecommunications companies that included Intel, IBM, Toshiba, Ericsson, and Nokia began to develop a way for users to connect a wide range of mobile devices without cables. This standard quickly gained new members including 3COM/Palm, Compaq, Dell, Lucent Technologies, and Motorola. Bluetooth membership numbered more than 2000 by year end 2000.

Bluetooth uses tiny radio transceivers that operate in the unlicensed 2.4-GHz band to transmit voice and data at up to 721 kbps. Bluetooth communicates within a 10-meter perimeter and does not require line of sight to establish a connection. According to Intel, the cost of the transceiver is about $20 and should eventually fall to $5.

Bluetooth will enable many new and important wireless data and control capabilities for wireless handsets. For instance, a document could be downloaded to a mobile handset and then printed on a Bluetooth-enabled printer, or devices in a home network could be controlled by the handset. Bluetooth can also be used to automatically synchronize PIM information between devices, when they are brought into range.

The ultimate value of Bluetooth will be determined by the number of Bluetooth-enabled devices available. Bluetooth will adhere to Metcalfe's Law, where the utility of the network is equal to the square of the number of nodes. Another example is the adoption pattern of the fax machine. Fax machines were not very valuable until there were a large number of fax machines with which to exchange faxes.

IDC expects Bluetooth volumes to ramp up very quickly, achieving penetration in more than 100 million devices per year by 2004. This is extraordinarily fast growth that may prove to be a bit optimistic, at this point, given the cost and timing of hardware cycles. However, once it becomes clear that there will be substantial actual demand versus forecast demand, manufacturers will get on board quickly. See Figure 3.12 for more details on Bluetooth-enabled hardware. Figure 3.13 shows the Bluetooth semiconductor market worldwide.

WAP Handsets

Much of the future in wireless communications will be driven by data communications. The first iteration of data/Internet-enabled handsets are Wireless Application Protocol (WAP) phones.

WAP is a global standard designed to make Internet services available to mobile users by converting Internet content to a format suitable for the limitations of mobile handsets, such as limited screen size and colors, lack of a keyboard or mouse, and limited memory and power. WAP is an enabling wireless Internet technology that uses

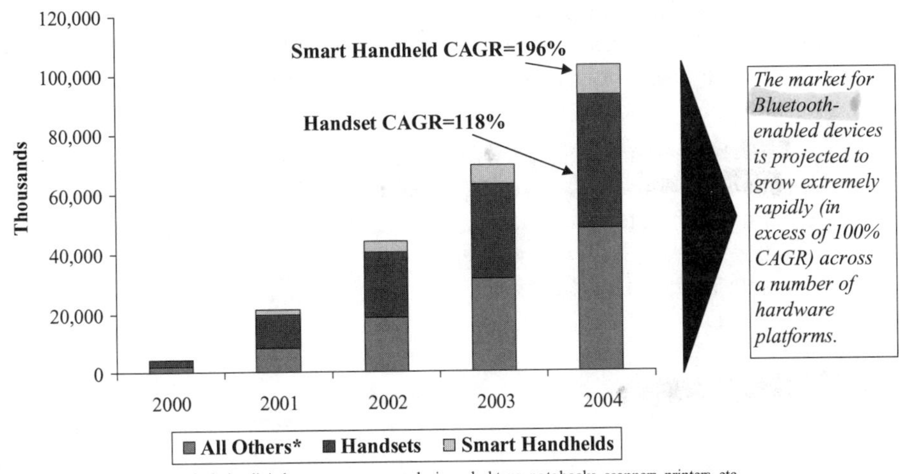

Source: IDC, 2000

Figure 3.12
U.S. Bluetooth Shipments by Hardware Segment.

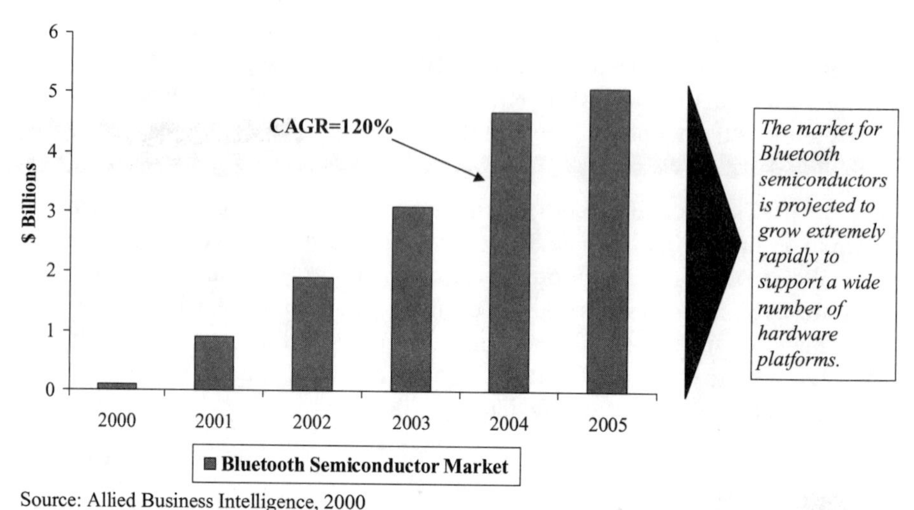

Source: Allied Business Intelligence, 2000

Figure 3.13
Bluetooth Semiconductor Market Worldwide.

wireless mark-up language (WML) and is more efficient at delivering information than HTML and similar languages.

WAP-enabled phones are expected to be a key driver in new handset sales and are expected to be supported by carriers who want to generate more revenue per user. Many experts predict that WAP phones or phones with browsers will represent 80–100% of all phones sold by 2002. This may, however, prove to be an overexuberant forecast as the business model for wireless data is hardly proven, particularly on a universal basis. Additionally, the current telephony handset is poorly designed to interact with data due to its small screen size and keyboard configuration. This may lead to a shift in form factor.

An alternative form factor, which may gain ascendancy with the advent of data, is the PDA, or Palm Pilot type device. This form factor is clearly better suited for interacting with data due to its larger screen (in some cases color), variety of input methods, greater memory, and greater processing power. PDAs are discussed more thoroughly in the PDA chapter.

GPRS Handsets

In 2000, only Motorola had a GPRS handset in production (see figure 3.14) and Nokia was planning to have their GPRS handset available by the second half of 2001. However, the initial GPRS handsets have not been able to achieve speeds as high as the advertised GPRS transmission rate of 115 kbps, due to issues with overheating.

Figure 3.14
Some Selected GPRS Handsets.

Smartphones

Smartphones are handsets that have more on-board processing power than a normal handset. They may also have PIM functionality. This can be enabled by having the Palm OS preloaded on the phone, like the Qualcomm pdQ. These phones were the first foray into a convergence product trying to marry the benefits of the PDA with the mobile wireless telephone.

As of January 1, 2001, Smartphones had not made a particularly positive impression on the U.S. market and volumes have been very low. Their sales have been restrained by cost, size, and slow data transmission speeds.

Structural Market Assessment

The mobile wireless handset market, despite its rapid growth, is not a structurally attractive market (see Figure 3.15). With a large number of competitors to choose from, and large make-or-break type orders, carriers have tremendous negotiating power and thus are able to extract significant value from manufacturers. Additionally, key component shortages routinely plague the industry. This is likely to continue as we move toward 3G as it is likely that 3G handsets will require expensive and hard-to-find color screens and powerful semiconductors.

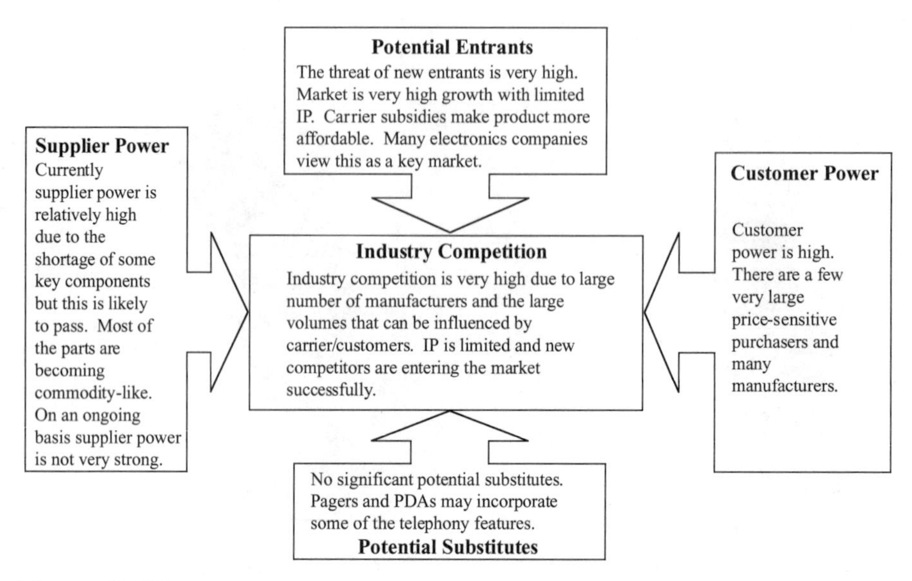

Figure 3.15
Structural Market Assessment: Handset Manufacturers.

Customer Power

Customer power is extremely high in the mobile wireless handset business. Carriers are extremely powerful and they control large volume orders. There are also a large number of capable suppliers to purchase handsets from, and once a manufacturer has designed a phone for a particular carrier's system, the investment must be recouped.

Potential Substitutes

There are no significant substitutes for the wireless handset.

Supplier Power

Supplier power has traditionally been relatively strong in this market, but on a rotational basis. At times, certain semiconductors, LCD screens, flash memories or other components are in extremely short supply. These types of shortages, over time, tend to remedy themselves in the handset manufacturers' favor. However, it is likely that a new shortage will crop up, as is potentially the case for color screens, which will favor specific suppliers for a certain period of time.

Potential Entrants

The mobile wireless handset market has a high potential for new entrants in the U.S. Handsets are an extremely large and high-growth market that many manufacturers view as key to their futures. There are also a large number of firms that sell mobile handsets outside of the U.S. that may want to enter the market. These firms include Sony, Philips, and Siemens.

The market structure also favors the entry of new manufacturers since the carriers can deliver a certain amount of volume to a particular handset if they desire. Additionally, the subsidy makes the product much more affordable to the end user, thus increasing demand beyond normal levels.

Industry Competition

Industry competition will be high given the number of suppliers and the power of the carriers. As would be expected in a structurally competitive market like handsets, margins are low with the exception of Nokia (Figure 3.16). Ericsson's handset business continues to deteriorate and, in January 2001, Ericsson announced that they were withdrawing from the manufacture of handsets and would subcontract with Flextronics for handset manufacture in the future, while still retaining development and design control.

Outlook

The outlook for mobile wireless handsets is one of very strong unit growth driven by both an increasing replacement cycle and solid growth in the number of subscribers.

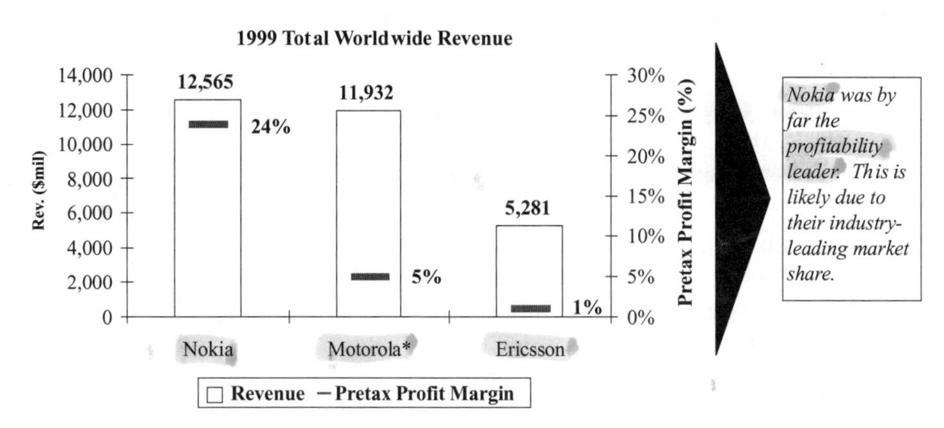

Figure 3.16
Wireless Handset Supplier Revenue and Margin Analysis.

Additionally, units should become more complex as expanded data and multimedia capabilities are enabled. This will entail greater processing power, attractive color screens, and integrated control functions such as Bluetooth. This will allow the mobile wireless handset to evolve into a substantial personal control, productivity, and entertainment device.

The complexity of supplying complex 3G capable devices may also provide temporary relief from the competitive forces that are roiling the handset market, as many suppliers may be unable to keep up technologically or generate a large enough financial return to justify the investment in 3G handsets. There are also likely to be shortages of key components that will limit overall handset availability in the earlier stages of 3G deployment, and data speeds are apt to be well below advertised levels.

The speed with which carriers deploy 3G networks will have a large impact on the handset market, as a quick roll-out of service will benefit the most capable handset suppliers and allow them to achieve substantial scale and experience advantages that will be difficult for second-tier firms to overcome. However, a slow roll-out may allow second-tier firms to participate in the market and gain experience over a longer lead time. The roll-out speed for 3G will, in large part, also be dependent on the availability of adequate handsets.

The shift to 3G should be viewed as a positive for handset makers as it allows them to have more intellectual property in the device, create more separation between first- and second-tier firms, and cause handset shortages. However, the investment will be substantial, particularly in the U.S. where there are likely to be a variety of standards in use. The variety of standards will also limit the available market for each handset. Substantial revenues from 3G handsets will probably not develop before 2004–2005.

PERSONAL DIGITAL ASSISTANTS (PDAS) . . . .

The Personal Digital Assistant (PDA), for the purpose of this text, refers to small hand-held devices that include processing and storage capability and generally run on the Palm Operating System (OS) or Windows Consumer Electronics (CE) software platform. The PDA includes a Personal Information Manager (PIM) and a variety of other applications, developed both by the manufacturer and by the developer community. In 2000, PDAs can have integrated wireless capability, an add-on wireless module, or no wireless connectivity at all.

The most popular PDAs are the Palm Pilot and Handspring lines, which both use the Palm OS (see Figure 3.17 and Table 3.4). Motorola, Qualcomm, Nokia, and Sony are notable firms that have licensed the Palm OS and are planning to introduce new products to the market. In the Windows CE camp, the most popular devices are the recently released Compaq Aero/iPaq and Hewlett Packard Jornada.

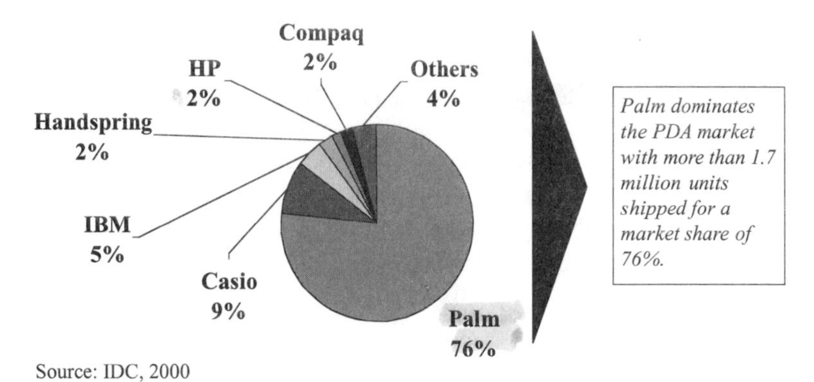

Figure 3.17
U.S. 1999 PDA Market Share.

Table 3.4 PDA Unit Market Share in U.S. Retail.

Manufacturer	Jan-00	Feb-00	Mar-00	Apr-00	May-00	Jun-00	Jul-00	Aug-00
Casio	7%	6%	3%	2%	3%	2%	1%	1%
Compaq	2%	1%	2%	1%	1%	1%	3%	7%
HP	2%	2%	1%	4%	6%	5%	5%	2%
Sharp	0%	0%	0%	0%	0%	2%	0%	0%
Other	4%	4%	1%	1%	1%	2%	1%	1%
Win CE Ttl	*14%*	*13%*	*7%*	*9%*	*12%*	*12%*	*10%*	*11%*
Handspring	0%	0%	1%	27%	31%	34%	36%	24%
Palm	86%	87%	91%	65%	57%	53%	53%	65%
Palm OS Ttl	*86%*	*87%*	*93%*	*91%*	*88%*	*88%*	*90%*	*89%*
Total	**100%**	**100%**	**100%**	**100%**	**100%**	**100%**	**100%**	**100%**

Source: PC Data

Market Size and Forecast

The market for PDAs, which is a relatively new category, has experienced tremendous growth, driven almost entirely by Palm. Palm introduced the first Palm Pilot in the spring of 1996. Palm currently has overwhelming market share, but many new competitors are appearing on the horizon.

The PDA is viewed as a key product by many manufacturers, especially as it gains wireless connectivity. Therefore, many new products are likely to emerge in this space (or adjacent spaces like Smartphones and RIM BlackBerry pagers). Firms from the consumer electronics, computer, and wireless communications industries are all likely to vie for supremacy in this market.

This category is poised to continue its explosive growth, and International Data Corporation (IDC) expects 33% compound annual growth in shipments through 2004 (see Figure 3.18). This would entail roughly $3 billion in U.S. market revenues. However, according to Salomon Smith Barney, these projected growth rates have already been eclipsed, and actual growth rates are likely to be twice as strong (Gardner 2000, 3). They base this belief on year-over-year sales growth rates in excess of 100% for Palm and Handspring (combined 1999–2000).

While extrapolating this type of growth too far into the future may be enthusiastic, it is clear that there is a tremendous opportunity for growth in this market. Growth can be expected to accelerate as the PDA gains increased functionality, such as broadband wireless access (network dependent), MP3 players, Bluetooth, and color screens.

Furthermore, the PDA market is far from saturated and is just beginning to take off. IDC estimates that, in 2000, the PDA had only achieved a 1.3% penetration rate. Comparing this to cellular phone penetration of close to 40%, it is easy to see why there is so much excitement in this product area.

Hardware

Palm dominated the market for PDAs in 1999 controlling three quarters of the market. However, with the advent of Handspring, a Palm OS licensee, the market has become more competitive. Handspring, which is a product of the original founders of Palm, has developed the attractive and innovative Visor line of PDAs. See Figure 3.19 for photos of the most popular PDAs.

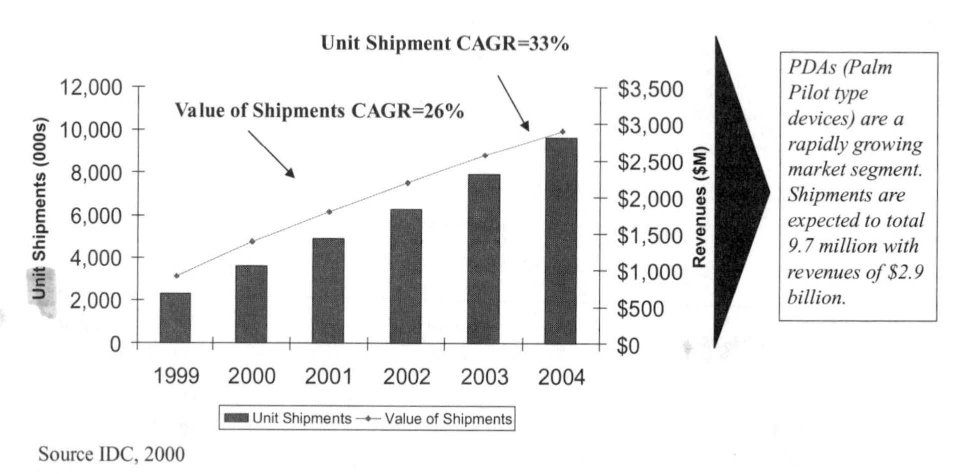

Source IDC, 2000

Figure 3.18
U.S. PDA Shipments and Revenues.

Palm™ m100

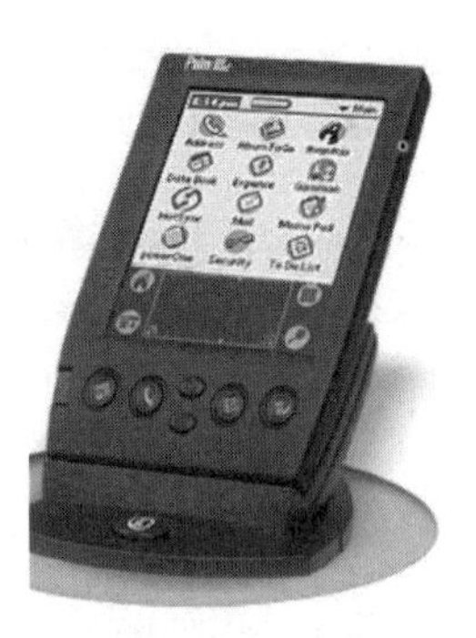

Palm™ IIIc

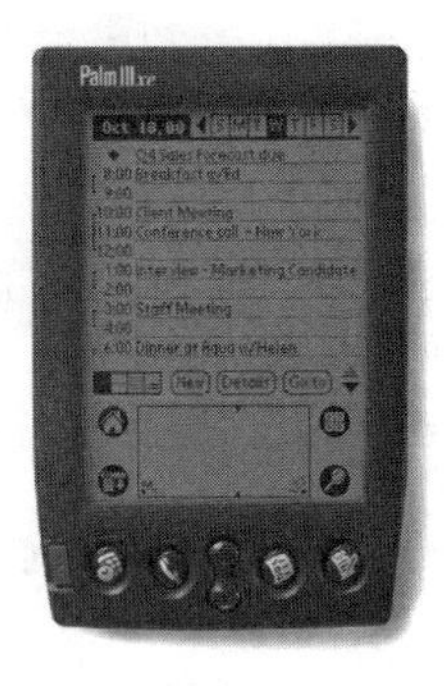

Palm™ IIIxe

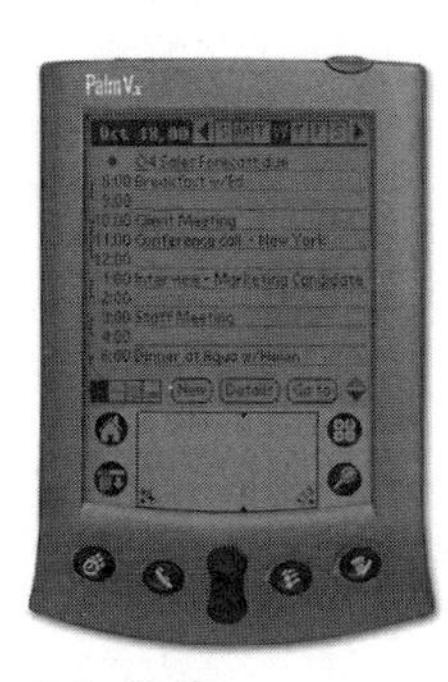

Palm™ Vx

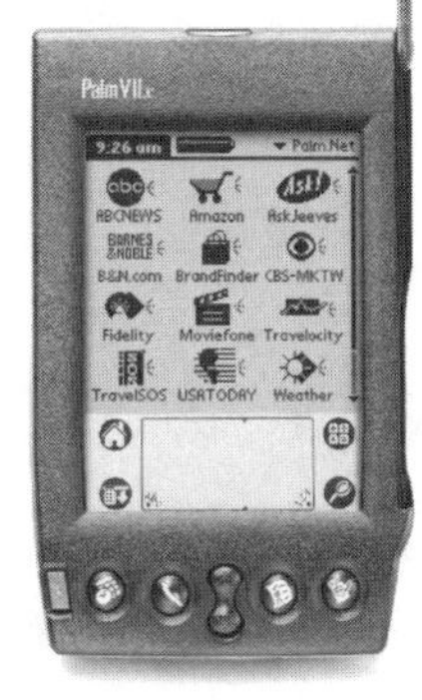

Palm™ VIIx

Source: Palm™ Web Site. Palm is a trademark of Palm, Inc.

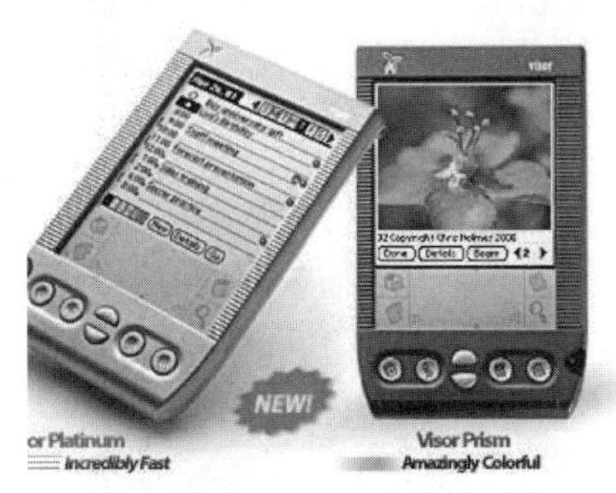

Handspring Platinum & Prism editions
Source: Handspring Web Site

Figure 3.19
Leading PDAs.

The key differentiating feature between the Visor and the Palm Pilot is the Springboard expansion slot. The Springboard allows the easy addition of hardware and software modules that autoinstall when inserted into the Springboard slot (see Table 3.5). Handspring develops its own modules and certifies modules from third-party developers. Handsprings are generally less expensive than comparably equipped Palm Pilots.

Operating Systems

The dominant operating system for PDAs is currently the Palm OS, although Microsoft's Windows CE is expected to gain a significant foothold (see Figures 3.20 and 3.21). Windows CE is expected to benefit from the strength of Microsoft's key manufacturing partners (Compaq and Hewlett Packard), its superior multimedia functionality, and its strong link to the desktop environment. Windows CE 3.0—introduced April 19, 2000—is also referred to as Pocket PC.

In comparing the two operating systems there are several points of differentiation that are worth mentioning.

Windows CE

The major advantages of the Windows CE (Win CE) Operating System, versus the Palm OS, are its greater multimedia capabilities and its greater interoperability with

Table 3.5 Handspring Modules.

Commercially Available Modules	Modules under Development
	Communications Modules
Modem	eBooks
8MB Flash Memory	Games
Tiger Woods PGA Golf	Scanners
Digital Camera	Sensors
Universal Remote	Recorders
Physician's Desk Reference	Remote Controls
Star Trek BookPak	MP3 Players
Keyboard	GPS
	Memory and Storage

Source: Handspring

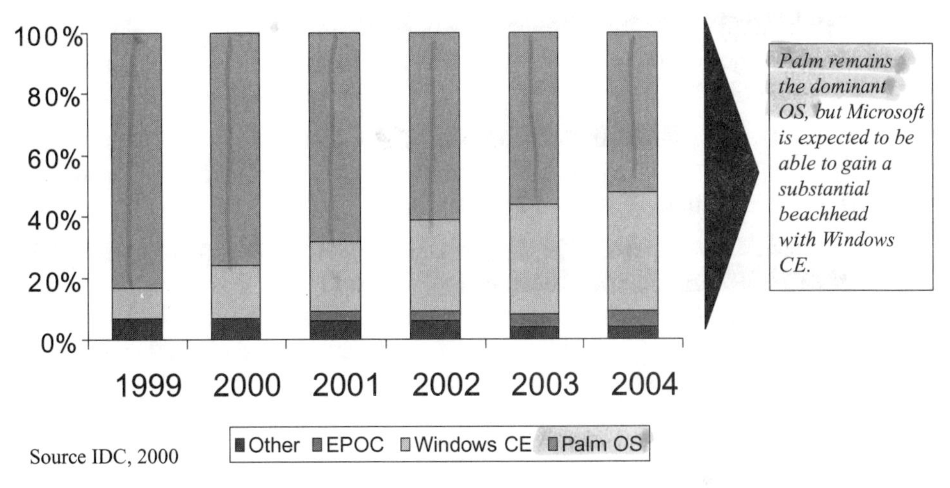

Source IDC, 2000

Figure 3.20
U.S. PDA Operating System by Share.

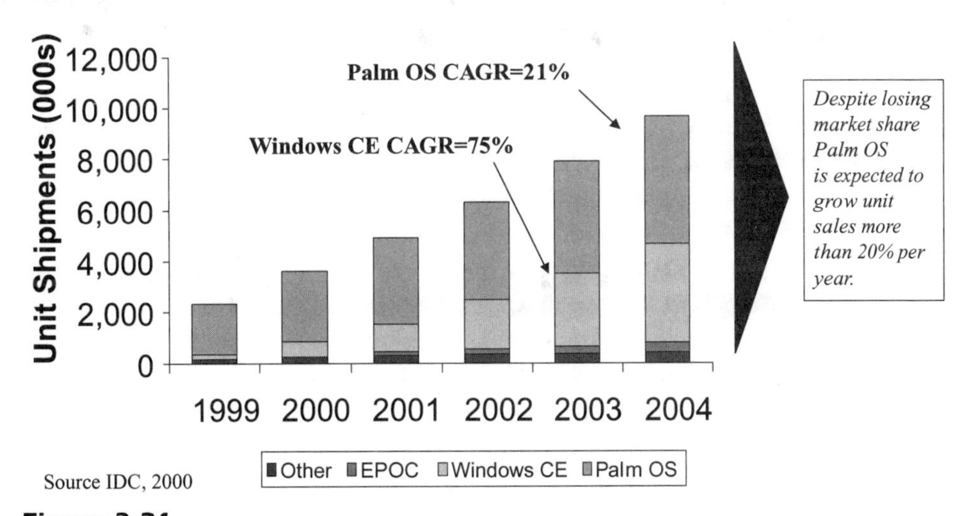

Source IDC, 2000

Figure 3.21
U.S. PDA Shipments by Operating System.

desktop programs. Windows CE comes "out of the box" with Pocket Word, Pocket Excel, Pocket Internet Explorer, Windows Media Player, and an e-Book Reader, in addition to the Outlook-based PIM.

Windows CE is also capable of displaying as many as 65,000 colors (256 for Palm), supports MP3 and full stereo sound, includes a voice recorder, and can process full-motion video. Win CE also has the ability to reformat and display HTML, eliminating the reauthoring of Web sites to WAP. This enhanced support for graphics and audio should enable Microsoft to gain an advantage in wireless multimedia applications.

The main disadvantages of Win CE are that it requires processors that severely limit battery life, it has a larger memory footprint, and previous versions of Win CE have been criticized as unstable (customers do not want to reboot their PDA).

It is important to note that, historically, Microsoft's third version of a program is usually the one that gains critical success. This has been the case with both Windows and Internet Explorer. The current edition of Windows CE is Microsoft's third version of a hand-held software platform and they have been very focused on making it a success.

Palm OS

The major advantages of the Palm OS, versus Windows CE, are its excellent power management capabilities, its simplicity, and its elegant user interface. The Palm OS benefits from a large library of shareware and commercial applications. It also benefits from a large and robust developer (more than 100,000 registered developers and growing rapidly in November 2000) and partner community, recently dubbed the "Palm economy" (see Table 3.6). Palm plans to make all of their future devices wireless capable and has publicly stated their firm commitment to wireless support for both their devices and the Palm OS.

The main disadvantages of the Palm OS are its lack of multimedia capabilities and its limited desktop application extension capability. Looking to the future, the conversion of the Palm OS to support multimedia functionality may be problematic, given the magnitude of the changes that need to be made.

Palm is making a concerted effort to make their operating system the *de facto* standard for all handhelds. To this end, they have adopted an aggressive licensing strategy which opens up the OS to a wide variety of hardware manufacturers (see Table 3.6). According to Salomon Smith Barney, the typical licensing agreement consists of an upfront fee, a percent of the licensee's quarterly revenue, and an annual support/maintenance fee. The range of per unit fees is 2–5%, with PDAs closer to 5% and phones closer to 2% (Gardner 2000, 33).

To further promote the Palm OS, on November 1, 2000, Palm announced a $100 million advertising campaign for the OS and its available applications.

Table 3.6 Palm Operating System Licensees.

Licensee	Product	Date of Agreement
Symbol	Handhelds	10/97
Qualcomm	Smartphone (sold business to Kyocera)	2/98
TRG	Handheld (in development)	10/98
Handspring	Visor Handheld	11/98
Nokia	Smartphone (in development)	10/99
Sony	Clie Handheld	11/99
Kyocera	Smartphone	9/00
Motorola	Smartphone	9/00

Source: Palm Web Site

Wireless

It is clear that the future of the PDA will be wireless (see Figure 3.22 and 3.23). The PDA has a number of key advantages over handsets in accessing, manipulating, and transmitting data in a mobile wireless environment. Chief among these are the larger screen size, local processing and storage, and a variety of input methods.

According to IDC, by 2003 almost 40% of PDAs will enjoy some form of wireless connectivity. There are two primary ways of wirelessly enabling a PDA—through

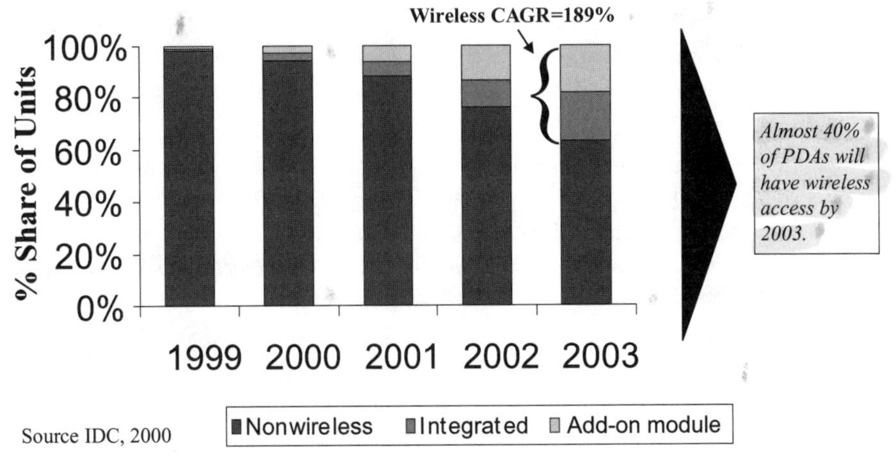

Figure 3.22
U.S. PDA Share of Units with Wireless Capability.

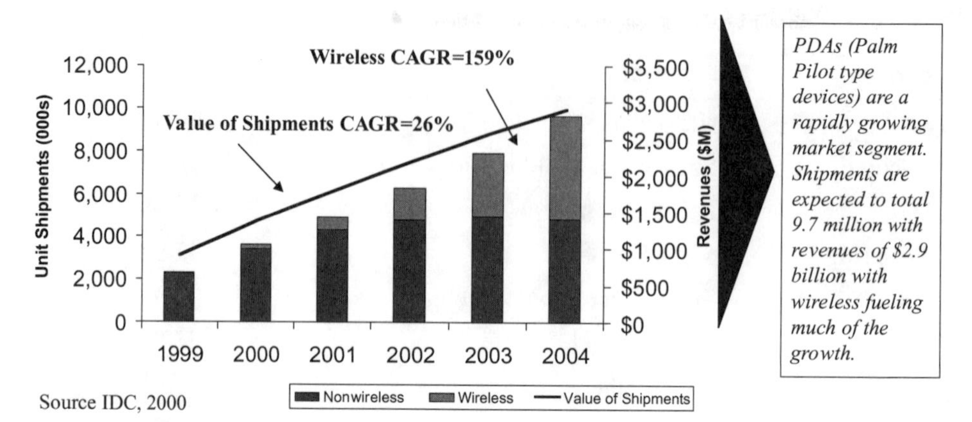

Figure 3.23
Wireless Component of U.S. PDA Shipments and Revenues.

an add-on wireless module or through integrated wireless capability. In 2000, the Palm VII was the only PDA with integrated wireless connectivity (also see RIM pagers, particularly the RIM 957 in the section on pagers [p. 75]). There are also attachable modems, or add-on modules, for the Palm V and Handspring PDAs, with planned support for more manufacturers. These add-on modules are sometimes referred to as sleds, as the PDA slips into the add-on's cradle. The dominant supplier of add-on modems has been Novatel. Omnisky introduced the Novatel Minstrel attachable wireless modem for the Palm V in May 2000; it supports the OmniSky wireless ISP service.

The Palm VII uses Web Clipping and can connect only to Palm's proprietary Palm.net service. Palm.net—launched in October 1999—runs exclusively on the BellSouth Wireless Data Mobitex network. The Mobitex network is a narrow band, packet-switched network with data transfer speeds of 8 kbps (very slow) for the Palm VII.

Web Clipping is a proprietary solution developed by Palm to display Web content on their PDA. Under this scheme, a template of the Web site is locally stored on the Palm device and only the requested information is updated. This procedure drastically cuts down on the amount of data transferred. Full Internet access is not available through Palm.net, but more than 400 Web sites were supported in 2000.

The OmniSky attachable modem connects to the Cellular Digital Packet Data (CDPD) network. CDPD has packet-switched data transfer rates as high as 19.2 kbps. Additionally, GoAmerica offers Internet access for a number of devices, including the Palm, on all of the major wireless networks including Mobitex, 2G cellular, and CDPD. There is further discussion of OmniSky and GoAmerica in Chapter 5.

To further enable their devices, both Handspring and Palm have announced wireless telephony modules for their products. Handspring has launched the VisorPhone

Table 3.7 Select Wireless Internet Access Plans for Handhelds.

Company	Network	Monthly Plan	Specifics
AT&T Wireless	CDPD 19 kbps	Local - $29.99	Unlimited wireless access on AT&T CDPD network; off-network $0.05 per kb.
GoAmerica	CDPD/BSWD 8–19 kpbs	Go.Lite - $9.95	Includes 25 kb of usage; additional usage is $0.10 per kb on CDPDand $0.30 per kb on BSWD
		Go.Unlimited - $59.95	Unlimited access
OmniSky	CDPD 19 kbps	$39.95	Unlimited access
Palm.net	BSWD 8 kbps	Basic - $9.99	Includes 50 kb of usage; additional usage is $0.20 per kb.
		Expanded - $24.99	Includes 150 kb of usage; additional usage is $0.20 per kb.
		Volume - $39.99	Includes 300 kb of usage; additional usage is $0.20 per kb.
		Unlimited - $44.99	Unlimited access
Verizon	CDPD 19 kbps	Basic - $24.95	Unlimited wireless access on Verizon CDPD network; off-network $0.08 per kb.
		Traveler - $24.95	Unlimited wireless access on Verizon CDPD network; off-network $0.06 per kb.

Source: W. R. Hambrecht, 2000

Springboard expansion module that operates on GSM 1900 MHz in the U.S., while Palm and RealVision have announced a GSM phone and Internet snap-on module for the Palm V in Europe and Asia. Additionally, a number of wireless handset manufacturers have licensed the Palm OS.

Overlaying IDC's wireless penetration figures onto their market growth projections, it appears that virtually all of the growth in the PDA segment will be wireless related. This may, in fact, prove to be conservative if broadband packet-switched networks are in general deployment before the end of the forecast period. Additionally, with the availability of attachable modems many of the existing PDAs can be wirelessly enabled.

When broadband packet-switched wireless networks are in general deployment, the PDA is likely to develop into a mobile multimedia entertainment device. This will have many implications for both businesses and consumers, as large amounts of data will be able to be transferred, enabling video and audio files to be transmitted wirelessly to the PDA. This will enable the PDA to evolve from a business productivity tool into a consumer entertainment device, as both the cost of hardware and transmission reach consumer mass market levels.

The PDA enjoys substantial advantages over the wireless handset in a data-centric world, and is likely to be the preferred form factor for interacting with data and video. The advantages of the PDA include larger screen size, greater processing power, greater storage capacity, and more user-friendly input methods.

Pricing

The price of PDAs, like most consumer electronics products, is expected to decline over time. However, the decline in the price of PDAs is expected to be more moderate than the decline in the price of handsets. This is due to the structural dynamics of the market and the addition of value-added features, such as color screens, wireless connectivity, and increased processing and storage capabilities (see Figure 3.24).

In September 2000, Morgan Stanley Dean Witter reported that Palm's average selling prices (ASP) fell to $240 from $262 in the previous quarter (Munson et al 2000, 3). This decline was attributed to the introduction of the $149 m100 model. Palm expressed the view that they expected ASPs to stabilize or increase in the future. Merrill Lynch estimates that Handspring's ASP for the third quarter of 2000 (fiscal 1Q) was $195 (excluding accessories), flat with the prior quarter (Crawford 2000, 2).

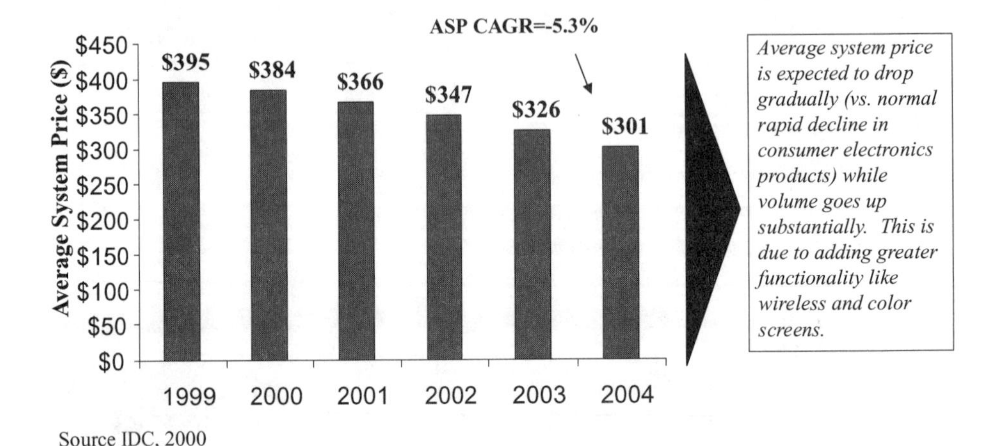

Source IDC, 2000

Figure 3.24
PDA Average System Price.

These prices are well below the IDC forecast of $384 per unit in 2000. However, the prices of Windows CE PDAs tend to be substantially higher than Palm OS products, costing more than $400 (see Figure 3.25). While the low end of the market has been expanded and become price competitive with the introduction of the Handspring line and the Palm m100, continued price firmness in the upper end of the market should be expected. ASPs of higher end units should continue to be driven up by improved functionality, like wireless connectivity, MP3 players, Bluetooth modules, and color screens.

Buyer Profile

Individuals are the primary purchasers (decision makers) in a PDA purchase, although more than one third of buyers are reimbursed by work. Table 3.8 shows the typical profile of these users, according to IDC.

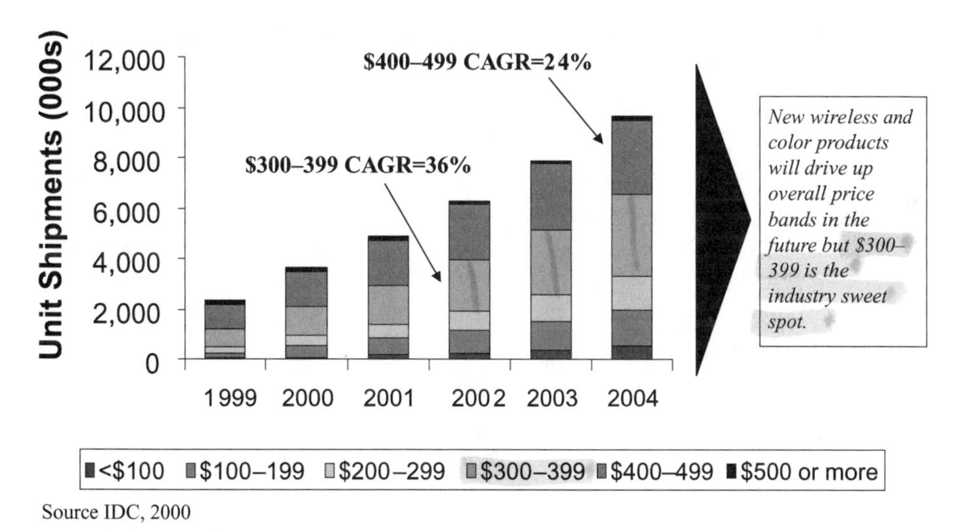

Source IDC, 2000

Figure 3.25
U.S. PDA Shipments by Price Band.

Table 3.8 U.S. PDA Buyer Profile.

Individual	95%
Male	92%
College Educated	84%
Uses for Business	75%
30–49 years old	70%

The uniformity of this profile is remarkable and suggests there is tremendous room for expansion in this category. For instance, achieving parity among females would virtually double the market.

Due to the fact that individuals are making these purchase decisions (see Figure 3.26) and many times are being reimbursed for the cost, the level of pricing pressure has been much less severe in the PDA market than in the wireless handset market.

To the benefit of the PDA manufacturer and industry pricing, the wireless carrier is not currently featured in the purchase decision. However, this may change over time, as data becomes a more important component of wireless traffic. When adequate bandwidth is available, the carriers are likely to identify the PDA as a large opportunity to drive additional traffic over their networks. PDAs' greater capabilities will enable providers to create sticky services aimed at retaining their customers. If the carriers begin to subsidize PDAs, as they have done with handsets, volumes will increase substantially and pricing will come under severe pressure. Therefore, it is imperative for PDA manufacturers to develop identifiable brands before the wireless carriers exert their influence over the market. Palm and Handspring have done excellent jobs of establishing their brands. See Figure 3.27 for a structural market assessment of PDAs.

In the meantime, corporate IT departments are most likely to become the primary volume purchasers of PDAs. Their motivation will be to increase the productivity of their mobile workforce, and their top priorities will be security and reliability, as it relates to accessing the corporate network. Price, while a consideration, will not be the number one concern for these purchasers. Additionally, controlling hundreds of, or a couple of thousand, orders will not give them enough leverage to dictate terms to manufacturers.

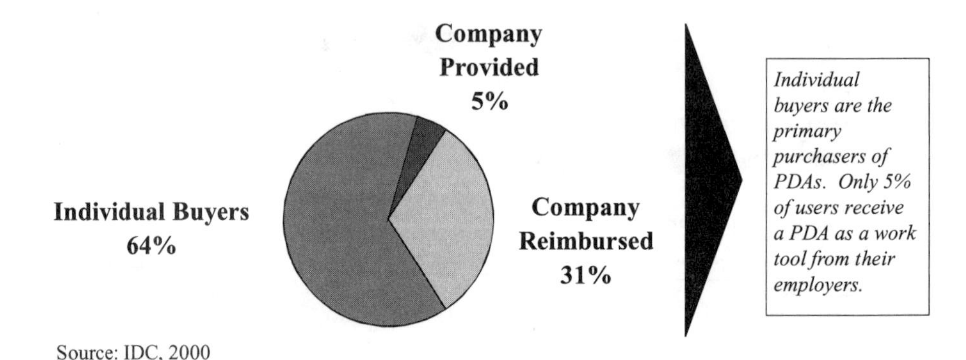

Source: IDC, 2000

Figure 3.26
U.S. PDA Purchasers.

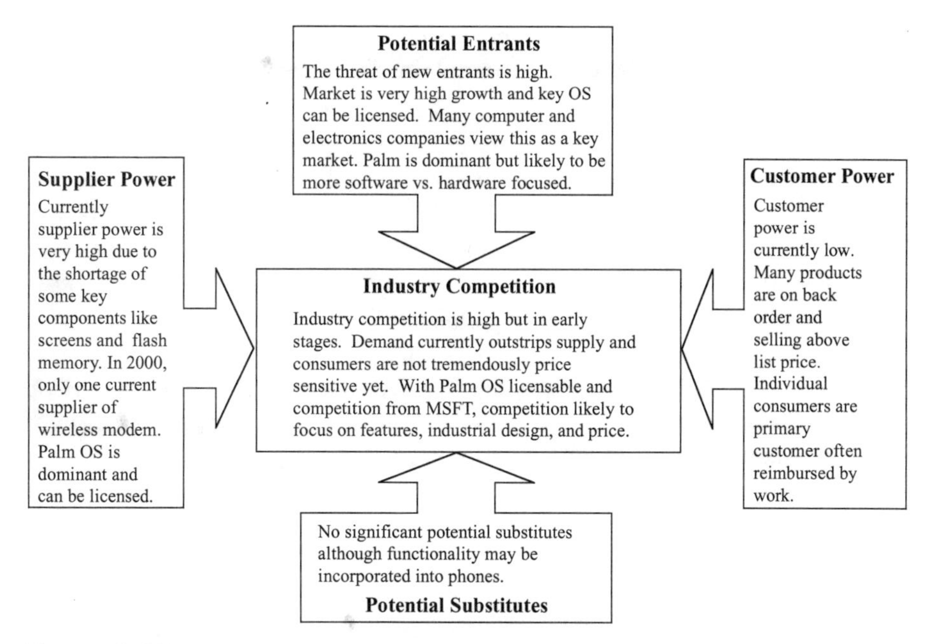

Figure 3.27
Structural Market Assessment: PDA Manufacturers.

Due to these favorable buyer dynamics, one can expect a positive, though competitive, pricing environment for PDA manufacturers. This is likely to remain the case until the wireless carriers begin to make a large push to enter the market.

Structural Market Assessment

The structure of the PDA market in 2000 is favorable, particularly with respect to the incumbents. Palm is presently in the driver's seat with overwhelming market and mind share, in both devices and operating systems. The decision to aggressively license their software should be viewed as a long-term positive for Palm. It increases the acceptance of their platform, from both a developer and a consumer perspective, and opens up a new stream of higher margin revenues. The move into OS licensing should also insulate Palm from falling manufacturing margins when wireless carriers and new competitors attempt to drive down prices.

Customer Power

Customer power in the PDA market in 2000 is particularly low. The market is essentially sold out, with most popular products on back order. In fact, some products are even selling above list price. There are no dominant customers in the market, and the typical consumer is an affluent business professional who is often reimbursed for the purchase by an employer.

The presence of a third-party payor in the market is a distinct positive for PDA manufacturers. A third-party payor generally leads to higher prices and greater functionality, as has been seen in the U.S. health care market. For PDAs, this will lead to integrated wireless capability, increased storage capacity, increased processing power, and color screens. Additionally, in terms of most customers' priorities, price tends to be relatively less important than functionality and ease of use.

Looking forward, the major threat of increased customer power is from the wireless carriers. When the wireless carriers decide to subsidize PDAs as wireless access devices, they can be expected to exert significant customer power. This will increase the volumes of the industry, but also reduce margins.

Potential Substitutes

The threat of potential substitutes is not high in a market growing this rapidly. While there may be other form factors that successfully emerge, it does not appear they will jeopardize the growth of the PDA market in the intermediate term. Likely substitutes in adjacent markets would include Smartphones and RIM BlackBerry pagers (although RIM 957 could be considered a PDA).

In the longer term, there is substantial expectation that there will be a convergence between the PDA and the handset markets, where the handset acquires PDA functionality and the PDA acquires telephony capabilities. There is already substantial evidence of this trend with telephony modules being introduced for PDAs and handsets being equipped with the Palm OS. However, there is enough demand for the individual products and form factors that two distinct markets are likely to remain, despite the availability of convergence products. Some customers will still appreciate the smaller size and lower price of the handset, and others will prefer the larger screen size and capabilities of the PDA.

Supplier Power

Suppliers are currently quite powerful—Palm in particular, as the supplier of the dominant OS. This situation is further complicated by Palm's position as the dominant PDA manufacturer. Thus, Palm competes with its OS customers. However, due to the popularity of the OS among consumers and the desire of many manufacturers to enter the market, Palm has been able to dictate terms to the industry. This may be an unsustainable position in the long run, and Palm may ultimately decide to exit the hardware business.

The presence and determination of Microsoft will serve to moderate Palm's power as an OS supplier. Microsoft has very deep pockets and has traditionally been a particularly fierce competitor, as the Justice Department has pointed out. Microsoft will continue to support and refine their product, and actively solicit manufacturers and developers. PocketPC may also prove to be superior to the Palm OS as a multimedia platform.

On the component side, in 2000 suppliers have significant power since many items are essentially sold out, and manufacturers are on allocation for key components like flash memory and LCD screens. However, over time, one would expect this supply situation to come into balance with PDA manufacturers gaining the upper hand.

Potential Entrants

The threat of new entrants into the PDA market is very high. This is an attractive high-growth, high-tech market that many industries can claim is core to their future. Entering the market is made easier by the ability to license the dominant OS on apparently reasonable terms.

Potential new entrants will emerge from the consumer electronics, computer, and wireless communications industries. This trend is already well underway, as Sony, Compaq, and Hewlett Packard have entered the market. This will intensify as these devices become predominantly wirelessly enabled.

Once broadband packet-switched networks are broadly deployed, the PDA will likely become the dominant mobile broadband multimedia terminal. This is a long way from its roots as an electronic organizer and places the PDA at the intersection of the mobile wireless communications, consumer electronics, and computer industries. Therefore, it should be expected that interest in entering this market will be particularly high.

Industry Competition

With these market dynamics, industry competition should be expected to be high, but not destructive. This market is in its early stages, and growth is expected to be rapid. The rapid growth of the industry should provide ample room for many competing ideas to be expressed in the marketplace. In fact, in 2000, the demand for PDAs outstripped supply. This may continue as PDAs evolve and gain more functionality through wireless connectivity, increased processing power, color screens, and innovative new features like MP3 players and Bluetooth modules.

Customer power is low and consumers in this market do not appear to be particularly price sensitive. This may be due to the fact that they are frequently reimbursed by their employer. Additionally, corporate IT departments may start to provide PDAs to employees to increase their productivity. Security and reliability are expected to be much higher priorities for these buyers than price.

With the ability to license the dominant OS, firms can compete on innovative industrial design, manufacturing capabilities, and the ability to integrate new value-added features. The presence of Microsoft, and their competing OS, should keep a lid on Palm's pricing and power. Microsoft is committed to this market and will compete vigorously, as they have always done. For evidence of this, one needs to look no further than the Desktop Applications and Web Browser markets, as Word Perfect, Lotus, and Netscape can attest.

In summary, this is a very attractive high growth industry that will continue to grow rapidly under a comparatively benign pricing environment (by consumer electronics standards). Industry competition can be expected to be high, particularly at the low end of the market, but is unlikely to be destructive in the medium term. The high level of growth and the need to integrate complicated new features, including wireless, should shelter the industry from runaway price competition.

Outlook

The outlook for the PDA market should be viewed as very favorable, and high levels of growth and rapidly increasing functionality should be expected to continue well into the future. The PDA market is also likely to be the most attractive of the terminal device markets.

As a relatively new category, PDA penetration is very low, leaving tremendous room for growth. Growth will also be fuelled by an increase of PDA functionality. As an electronic organizer, PDA growth is limited; however, as a mobile multimedia terminal, the future of the PDA is extraordinarily bright. Due to its substantial advantages, the PDA is likely to be viewed as a superior form factor for interacting with wireless data. If the PDA is able to secure only a small portion of the wireless handset market, expected to be close to 80 million units in the U.S. for 2001, the category's growth is assured.

The extremely positive outlook for this industry will attract firms and encourage them to present innovative new products with ever-increasing capabilities to the market. The PDA market will be tremendously exciting well into the future.

PAGERS

Pagers are devices used to receive paging network signals. The first paging systems alerted customers to an incoming message through a tone emitted by the pager. The subscriber then called in to an operator to retrieve the message. Numeric paging was then developed, and the subscriber was alerted through a number that appeared on a small LCD display on the pager. This was generally a phone number. Alphanumeric paging has enabled the subscriber to receive text information on pagers with displays of various sizes.

Traditionally, paging has been a one-way medium. However, with Mobitex, ARDIS, and the advent of ReFLEX, interactive mobile data communication has been enabled. This allows customers to send and receive email through their pagers, using a QWERTY keyboard.

Pager Market Size

Consistent with the growth of interactive paging, equipment sales will migrate rapidly from numeric to alphanumeric products. Strategy Analytics predicts that the total number of pagers will remain relatively flat, in the 25 million per year range (see Figure 3.28). This represents a large market in terms of unit volumes for equipment suppliers. However, as with most electronic products, pricing is expected to erode.

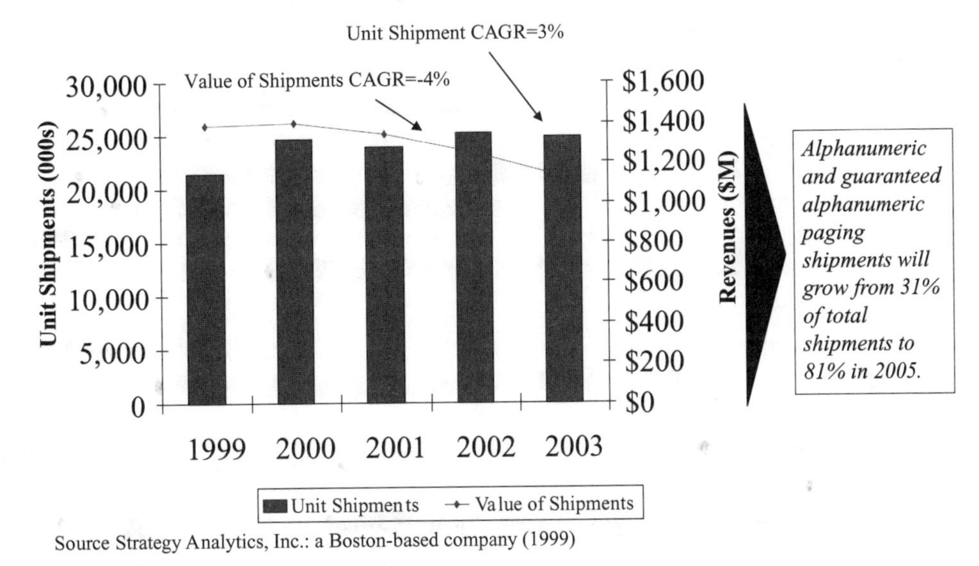

Figure 3.28
U.S. Pager Shipments and Revenues.

Total revenues are projected to decline from $1.4 billion to about $1.1 billion. This translates to an average price of about $40 per pager.

Pager Customers

Historically, a large number of pagers have been purchased by the network operators and then leased to customers. This practice led to large capital expenditures and depreciation charges for the carriers. The current financial status of the industry, and consequent rationing of capital, has made this a much more difficult strategy for the carriers to pursue. As a result, paging companies are encouraging customers to purchase the units themselves. This generally requires a subsidy and leads to a marginal loss (revenue less than cost of goods sold) on every unit sold by the carrier.

Pager Suppliers

Motorola is the dominant supplier of pagers in the U.S., with share estimates above 70% and 2000 sales of about a billion dollars. They produce a full line of pagers, including a variety of stylish interactive pagers that came to market in the latter half of 2000 (see Figure 3.29). Glenayre Technologies is also active in the pager market and has taken over Motorola's infrastructure business through a licensing agreement.

One of the most innovative and successful two-way messaging or paging products has been the RIM BlackBerry pager. The BlackBerry is a complete solution that provides hardware, software, and bundled airtime.

The BlackBerry, introduced in 1999, offers a synchronized (with the desktop) email address and push technology. Email is pushed to the BlackBerry over an always-on packet-switched network connection. This connection is secured using Triple DES encryption technology. The BlackBerry also has a QWERTY keyboard, an Intel 386 processor, and up to 5 MB of memory, making it a powerful device. The BlackBerry was the first product to provide a wireless synchronized email box, where the email address of the BlackBerry is the same as the desktop. This was a tremendous breakthrough.

The BlackBerry 850 operates on the Motient (ARDIS) network. The BlackBerry 950 and 957 use the BellSouth Wireless Data Mobitex network. The 850 and 950 are virtually identical.

The 957 adds many of the PIM functions typically associated with the PDA market. Given PIM functionality, a QWERTY keyboard, and the synchronized email address, the 957 is a strong PDA offering for the mobile professional. In comparison, the Palm VII requires a connection request, stylus input, and a separate email address, despite working on the same Mobitex network. This makes the 957 a uniquely capable product.

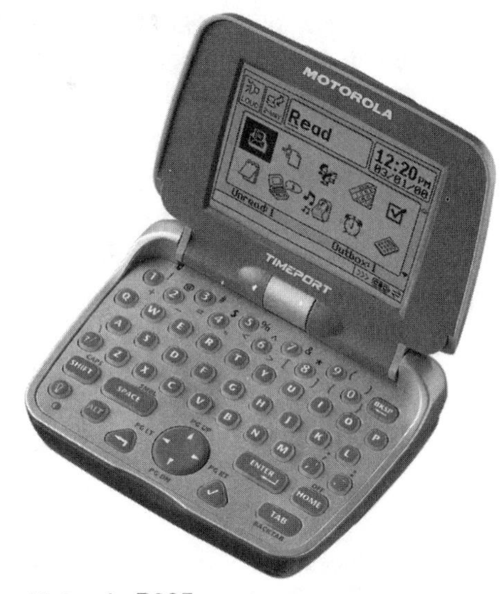

Motorola T900

Motorola P935

Source: Motorola. Reproduced with permission from Motorola, Inc. © 2001, Motorola, Inc.

BlackBerry (RIM 950)

BlackBerry (RIM 957)

Source: Research In Motion. The BlackBerry and RIM families of related marks, images, and symbols are the exclusive properties of, and trademarks of Research in Motion Limited. Used by permission.

Figure 3.29
Leading WIreless Handhelds

Traditionally, the BlackBerry products have been targeted at the enterprise market and are compatible with Microsoft Exchange and Outlook. A new version will be compatible with Lotus Notes.

The pager is sold directly by Research in Motion, and resold by value-added resellers (VARs), Compaq, Dell, Earthlink, Aether, Motient, and BellSouth Wireless Data. AOL has also signed an agreement to sell the BlackBerry for their Instant Messaging system, which targets consumers. As of August 2000, there were 75,000 BlackBerry subscribers, and the forecast was to add 25,000 more in the November 2000 quarter.

Service cost from RIM is generally $400–500 for the unit, and a fixed $40/month for airtime. Other vendors offer a variety of different plans—BellSouth Wireless Data offers the device for $369 plus $9.95–69.95 per month for airtime. With the exception of the AOL deal, RIM gets an estimated 40% share of the airtime revenue (Labe 2000, 4).

As is typical with pagers, the BlackBerry offers broad coverage, superior in-building penetration, and extended battery life. When 2.5G packet-switched networks come on line, voice capability may well be added to the service.

Structural Market Assessment

The pager market is probably the least attractive of the wireless terminal device markets. The entire industry is under siege from cellular networks, and this is unlikely to abate in the foreseeable future (see Figure 3.30). The future of the pager is interactive and there has been notable success in this product area from both Research in Motion and Motorola.

Customer Power

Customer power in the paging market is moderate, due to the limited number of suppliers and Motorola having more than 70% market share. Paging network service providers have been forced to shift their business model from purchasing and leasing pagers to having the subscriber purchase the unit. This has made the purchase decision an individual decision and reduced customer power.

The biggest threat from customers is that they abandon paging service in favor of cellular telephony. This helps to keep a lid on pager prices.

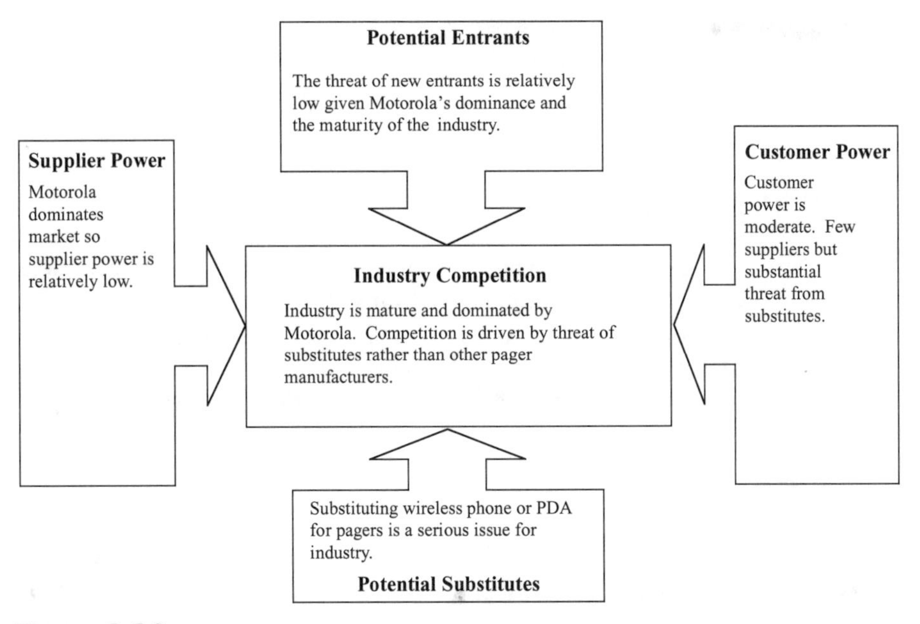

Figure 3.30
Structural Market Assessment: Pager Manufacturers.

Potential Substitutes

Potential substitutes are the biggest nemesis to the pager manufacturers. A wireless phone or wireless PDA can provide virtually all of the functionality of the pager, plus substantial additional capabilities. The price differential between a paging device plus service and a cellular handset plus telephony service is no longer sufficient to favor pagers. This is particularly true when cellular handsets are subsidized by the carrier, as is generally the case.

Supplier Power

Supplier power is moderate. Motorola, as the dominant pager manufacturer, is in a position to dictate terms to suppliers. Any component shortage should be regarded as temporary, with the manufacturer gaining the upper hand over time.

Potential Entrants

The threat of new entrants is moderate. The industry is in difficult shape and Motorola is far and away the dominant supplier. Any growth in the industry will be in interactive paging. Interactive pagers may tempt some manufacturers to enter the industry. However, it is unlikely that a large number of firms will find this market sufficiently attractive to enter.

Industry Competition

Industry competition among pager manufacturers is unlikely to be particularly fierce. Rather, it is the threat from potential substitutes that will drive the competitive intensity of the industry. Paging has been plagued by wireless telephony for some time and this should be expected to intensify. With the paging service providers in very difficult financial condition, it will be difficult for the pager suppliers to prosper.

Outlook

Pager manufacturers are benefiting from the introduction of two-way pagers. Motorola and Research in Motion have experienced substantial success with their interactive products. However, the desperate financial condition of the paging industry and the pervasive threat from cellular telephony will continue to cast a pall over the industry. The future of paging may lie in telemetry, and pager suppliers may well want to investigate this product area for future growth.

WIRELESS NETWORK INFRASTRUCTURE . . . .

There are two components of the wireless network infrastructure addressed in this section—infrastructure hardware and location technologies.

The wireless telephony network infrastructure hardware is the equipment that enables wireless communication between mobile terminals and other mobile or landline terminals. This infrastructure consists of several key elements—the switching system, the base station system, and the operation and support system.

Location technology is the ability to identify the location of a mobile terminal, and hence the subscriber, with a reasonable degree of accuracy, as it moves throughout the mobile wireless network coverage area. This capability is enabled through the network, the handset, or a combination of both.

Infrastructure hardware is one of the most structurally attractive segments of the wireless industry and is projected to grow rapidly and profitably with the introduction of 2.5G and 3G services.

Wireless Network Infrastructure Hardware and Software

Completing a wireless telephone call involves the following steps: the mobile handset communicates with a cellular base station that relays the signal to a Mobile Switching Center (MSC). If the caller is trying to reach another cellular handset, the MSC then routes the call to the appropriate base station where the receiver's handset is active. This base station then transmits the signal to the receiver's handset, completing the call. When communicating with a landline telephone, the Mobile Switching Center provides the interface to the Public Switched Telephone Network (PSTN), which then routes the call to the appropriate landline phone. Many mobile-to-mobile calls are also routed over the PSTN when a different MSC needs to be accessed.

Wireless telephony does not function through handset-to-handset communication, like a walkie-talkie.

The wireless telephony network is structured as a patchwork of hexagonal cells, like a honeycomb (see Figure 3.31). Each cell contains its own radio equipment and is allocated its own set of voice channels. Adjacent cells are assigned different channels in order to avoid interference. Nonadjacent cells, however, are able to reuse the original channels, thus increasing capacity. As the network gains more subscribers, the power of each transmitter can be reduced so that cells can be located closer together, reusing the spectrum more frequently.

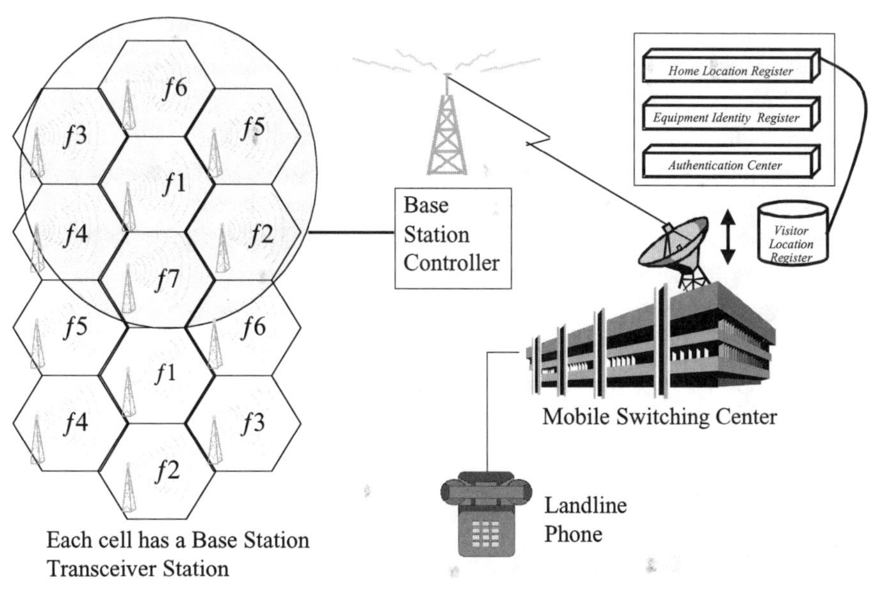

Figure 3.31
Cellular System Architecture.

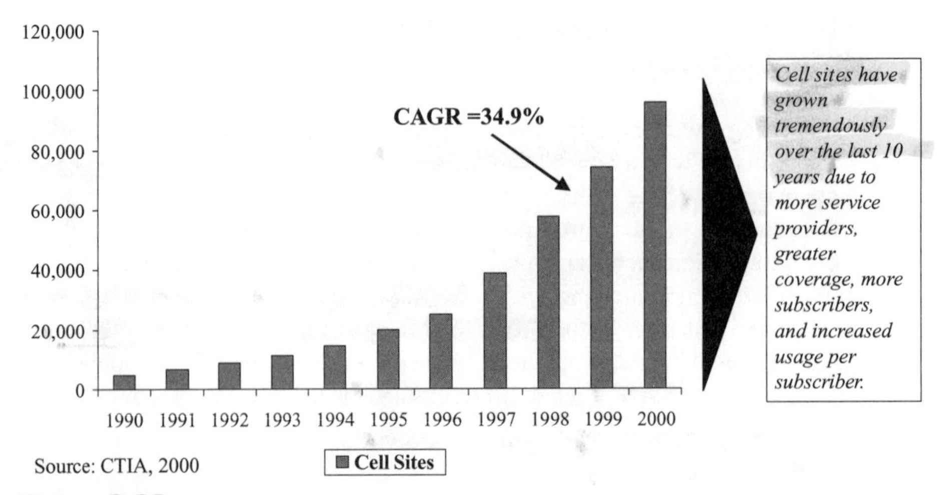

Figure 3.32
U.S. Cell Sites in Commercial Use.

Base Station System

The Base Station System provides the interface between the mobile handset and the wireless network. This system is the most costly part of the wireless infrastructure and can represent about 70% of infrastructure hardware cost.

The main components of the Base Station System are the Base Station Controller (BSC) and the Base Station Transceiver Station (BTS).

The Base Station Controller is the hub of the wireless infrastructure, handling all connections with moving mobile terminals. The BSC provides the connection between the mobile system and the MSC. The BSC is responsible for channel setup, frequency hopping, and call handoffs (switching channels in the same cell and switching cells under the domain of that particular BSC). The Base Station Controller can connect as many as several hundred Base Station Transceivers.

Call handoffs, which transfer the call from one BTS to another, can occur at the request of the network—in order to balance network traffic—or at the request of the handset, when signal quality is poor. Under GSM, the handset scans for the six best nearby cells for a transfer and relays this information once per second to the network. There are a number of algorithms that determine when a handoff should be made.

The Base Station Transceiver Station transmits and receives signals to and from the mobile terminal, providing the interface to network (see Figures 3.33 and 3.34). The BTS is basically the antenna and it handles the bulk of the radio features of the wireless infrastructure. Since transmission of the radio signal is proportional to the height of the antenna, you will often see cellular antennas (BTSs) on hilltops or the top of tall buildings.

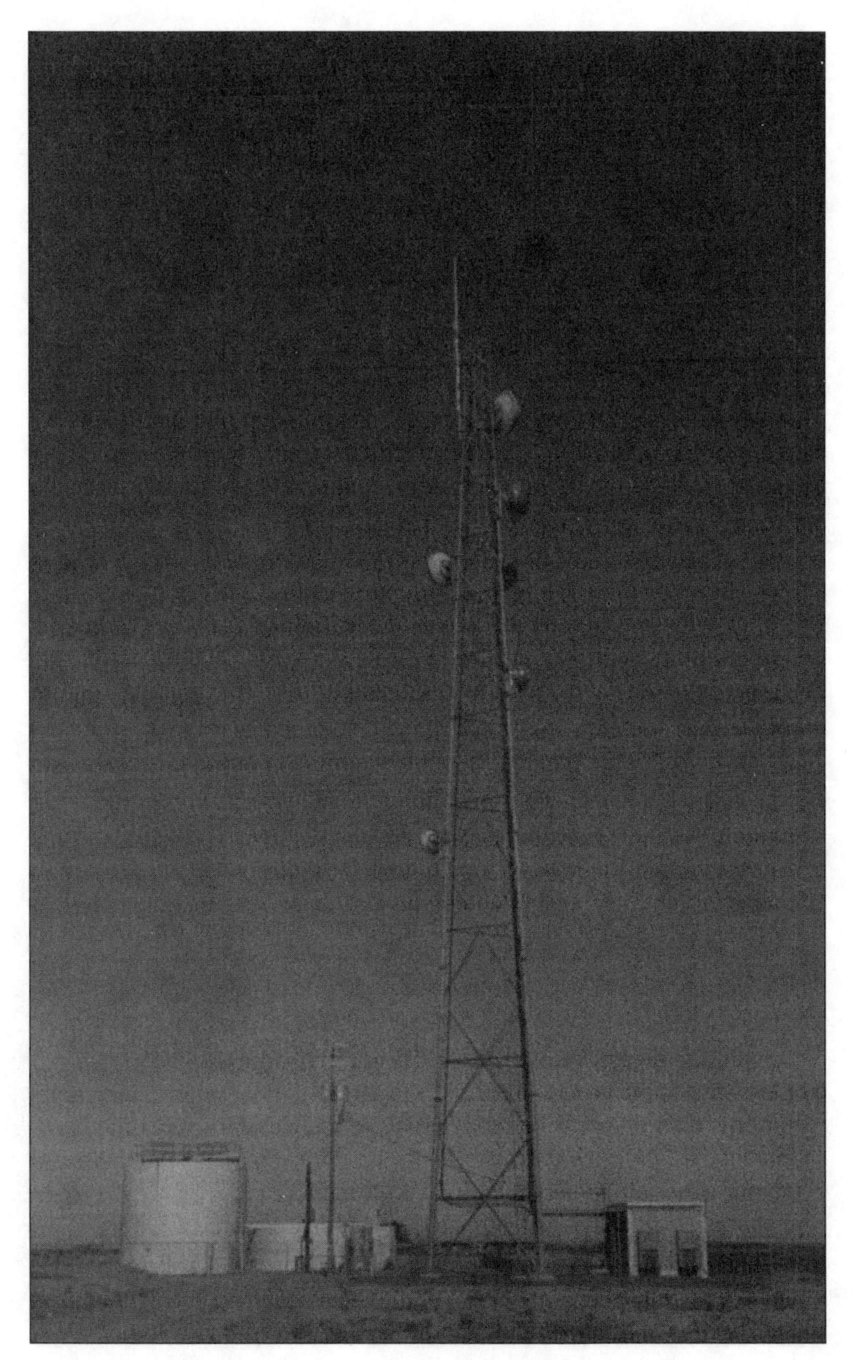

Figure 3.33
Self-Support Tower in California. Courtesy of American Tower Corporation.

Figure 3.33
Monopole with Base Station Equipment. Courtesy of American Tower Corporation.

Mobile Switching System

The Switching System performs the switching, routing, and call control functions. It also handles billing, call services, and intelligent networking. The key components of the switching system are the Mobile Switching Center, the Home Location Register, the Visitor Location Register, the Authentication Center, and the Equipment Identity Register.

The Mobile Switching Center (MSC) provides the traditional switching function of the landline phone network. This function is to open a dedicated circuit between the caller and the receiver. The major suppliers base their mobile switch on their landline switch. Thus, Lucent bases their mobile switch on their 5ESS switch. The MSC provides the connection to the Base Station System and to the landline network and sets up calls to other MSCs. The MSC also handles call handoffs not handled by the BSC (calls under different BSCs but the same MSC and calls under different MSCs). Each MSC can handle about 100,000 to150,000 subscribers.

The Home Location Register (HLR) is a centralized database that stores information on all subscribers. The HLR also maintains location information about the subscriber.

The Visitor Location Register (VLR) is a database that keeps a record of all mobile subscribers currently active in a particular MSC. The handset routinely sends signals to the VLR to alert the system to its presence. The VLR forwards this information back to the HLR so that calls can be properly routed to the handset.

The Authentication Center (AuC) is a database that keeps the authentication register of all subscribers. The mobile handset contains a key that must be authorized by the AuC for the handset to gain access to the network.

The Equipment Identity Register (EIR) contains information about all valid mobile terminals on the network.

Under the GSM system, the HLR, AuC, and EIR are usually co-located centrally, while VLRs are usually distributed throughout the network, generally with the MSCs.

The switching system can make up about 30% of the infrastructure hardware cost.

Operation and Support System

The Operation and Support System manages the network. This system handles system and cell planning, generates traffic reports, and measures radio signal, in order to optimize network performance.

In an infrastructure installation, hardware typically makes up 70–75% of the cost.

Infrastructure Suppliers

The main suppliers in the U.S. wireless infrastructure market are Ericsson, Lucent, Nortel, and Motorola. Their shares are outlined in Figures 3.35–3.39, and vary significantly by technology. Worldwide in 1999, Ericsson and Lucent combined to secure more than 50% of the infrastructure orders, with Ericsson the clear leader in wireless infrastructure. However, in CDMA technology, Lucent and Nortel are the leaders.

Market Size and Share

The IDC forecast for infrastructure spending in the U.S. market is relatively flat at about $3 billion per annum (see Figure 3.40). However, this is expected to pick up substantially when 3G services begin to be deployed. In the meantime, only 2G digital technology is projected to be deployed in the U.S.

The IDC forecast, however, does not agree completely with industry data gathered by the Cellular Telephone Industry Association (CTIA), which suggests capital expenditures for the industry from June 1999 to June 2000 were about $10 billion (see Figure 3.41). This is also consistent with Merrill Lynch data on contract awards that suggest a spending level at about $8 billion per year (this would imply more money actually spent since continuing expenditures are made outside of announced contract awards). The market for wireless communications has historically grown much faster than the most optimistic forecasts.

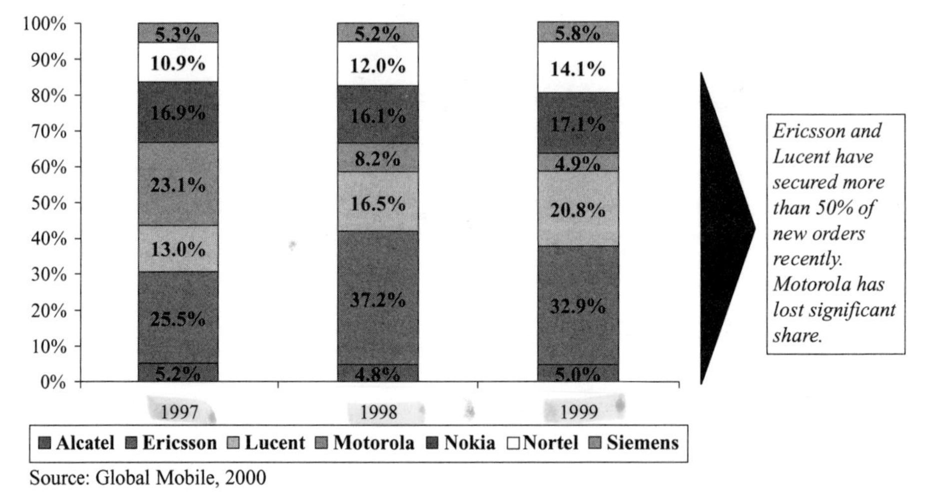

Figure 3.35
Wireless Infrastructure Percent of Orders (Global).

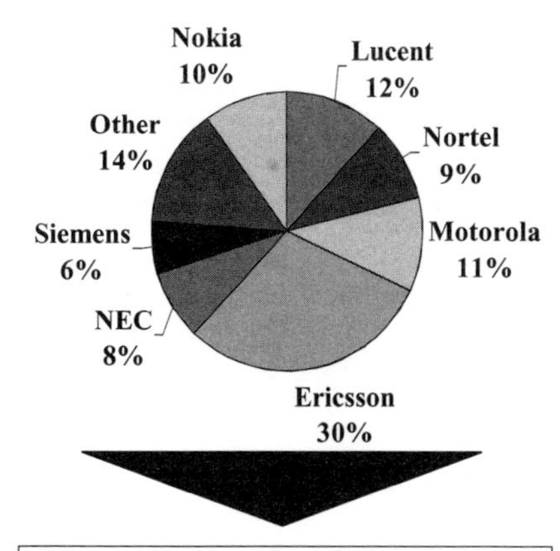

Figure 3.36
Global Wireless Infrastructure Market Share 1999.

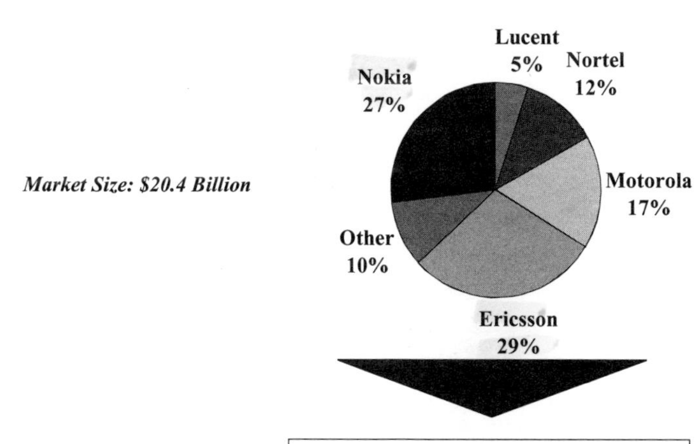

Figure 3.37
Global Wireless Infrastructure 2000 Market Share of Contracts (Value): GSM.

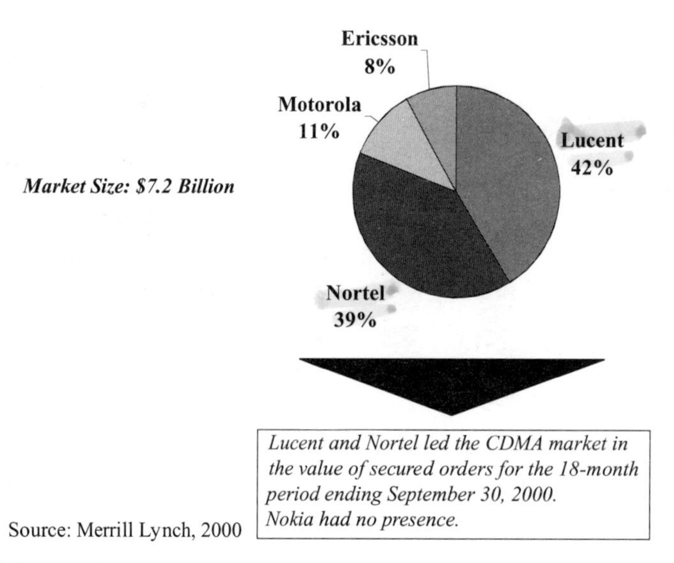

Figure 3.38
Global Wireless Infrastructure 2000 Market Share of Contracts (Value): CDMA.

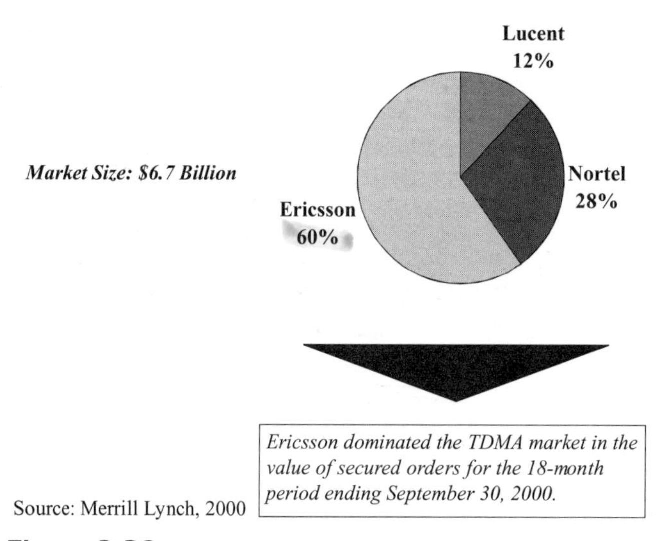

Figure 3.39
Global Wireless Infrastructure 2000 Market Share of Contracts (Value): TDMA

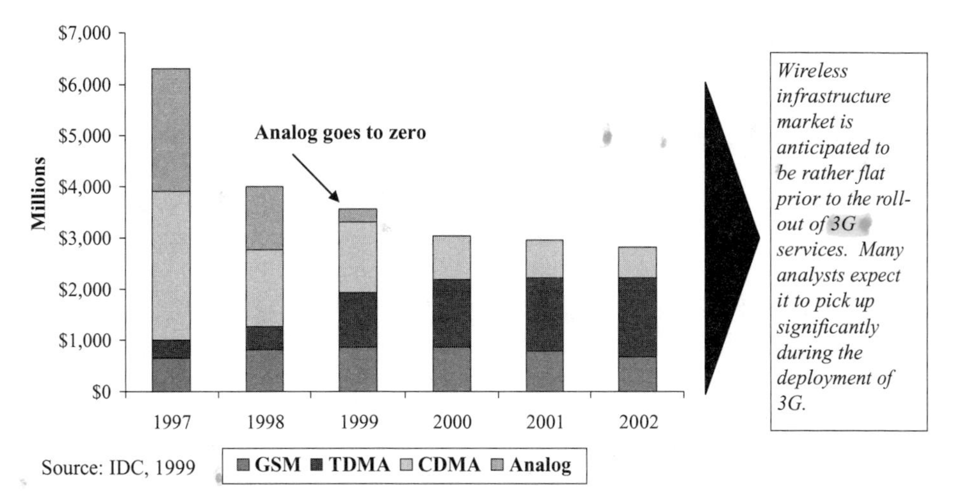

Figure 3.40
U.S. Mobile Wireless Infrastructure Spending Forecast.

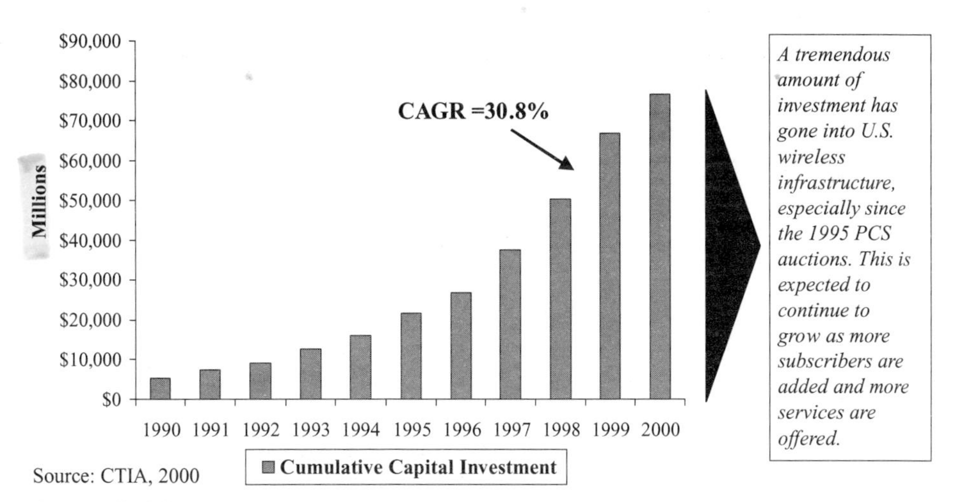

Figure 3.41
U.S. Cumulative Wireless Capital Investment.

Third-Party Communication Tower Operators

A recent trend in the wireless communications infrastructure market has been outsourcing. In 2000, the main item that was being outsourced was the communication tower. The communication tower holds the carrier's radio antenna, or base station. Some carriers view the communication tower business as noncore to their operations and have sold their towers to third parties. Towers are also very capital and maintenance intensive.

Third-party purchasers have been interested in purchasing the communication towers since they are able to lease space on the tower to more than one tenant. This can substantially increase their returns, especially since they are usually guaranteed an anchor tenant for a number of years as part of the purchase agreement. The major third-party communication tower owners in the U.S. are American Tower, Crown Castle, Spectrasite, SBA Communications, and Pinnacle (see Figure 3.42). The rent per tenant for space on a tower averages about $1,500 per month, but can vary widely based on location.

Between 1998 and 2000 many major carriers sold all, or a portion, of their portfolio of communication towers. The major carriers and the number of towers they divested included Bell Atlantic (1,427), Nextel (2,000), BellSouth (1,850), Airtouch

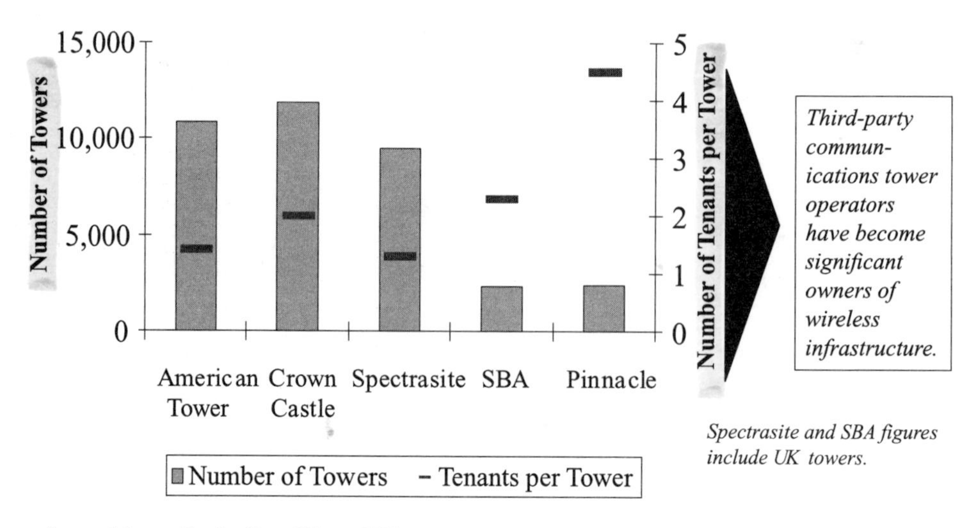

Figure 3.42
Communication Tower Operators' Number of Towers and Tenants per Tower.

(2,100), GTE (2,300), and Alltel (2,193). The prices received per tower varied between $262,000 and $423,000, with the average being about $350,000. Location, population density, and lease-back terms (including building additional new towers for the carrier) were the main contributors to price differences.

One of the major side effects of this practice is to make it much easier for firms to establish or expand service in an area where they have spectrum. This is because it is difficult and time consuming to obtain approval to build a tower, while adding an antenna to an existing tower is much less of an issue. This may be a substantial benefit when carriers move to 3G technologies that require more antennas for a given area. However, this can also increase the competitive intensity in an area.

Given the attractive prices being paid to carriers for their towers, their noncore nature, and capital intensity, this trend should be expected to continue.

Addition of Data and 3G

Paths to 3G

The addition of data and the move to 3G should drive substantial demand for wireless infrastructure equipment. Construction of 3G networks is expected to be exceedingly costly, with costs of infrastructure approaching the license costs. Enabling a nationwide 3G network will be a tremendously expensive undertaking, running into the many billions of dollars ($150 billion worldwide over the next five years according to S. G. Cowen). S. G. Cowen projects that the worldwide infrastructure market, driven by 3G, will grow at 22–25% over the next three to five years (Cowen 2000, 4).

Vodafone has reportedly signed a £4 billion (about $5.7 billion) infrastructure contract with Ericsson after spending £6 billion for its UK license, while Mannesmann has committed to spending 10 billion Deutschemarks (about $4.6 billion) to upgrade its network to 3G, after spending 16 billion Deutschemarks for its German license. It is important to note that these are just the preliminary contracts, and actual build-out costs are likely to be much higher. This points to a robust environment for infrastructure spending as new services are introduced and new subscribers are added. It also appears that minutes of use per subscriber are on the rise, which drives further capacity requirements.

Ericsson and Nokia have gained the early lead in GPRS and W-CDMA contracts, based largely on their strong positions in GSM (see Figures 3.43 and 3.44).

To enable data service, a network operator will have to add data servers or WAP gateways (see Figure 3.45). By the end of 2003, we may see data transfer speeds between 384 kbps and 2 Mbps. This is as fast as or faster than current high-speed DSL connections. Data will clearly be the long-term driver of the wireless industry and in

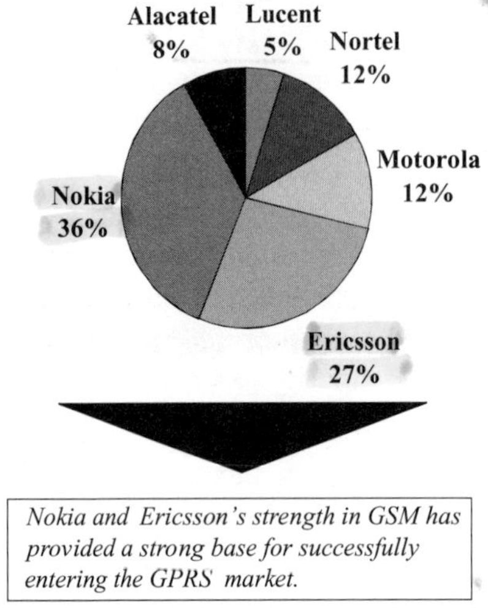

Source: S. G. Cowen, 2000

Figure 3.43
Global Wireless Infrastructure: Cumulative Announced GPRS Contracts.

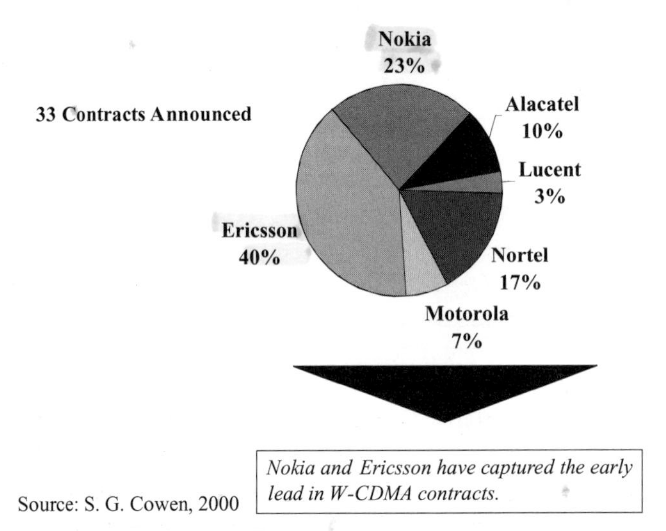

Source: S. G. Cowen, 2000

Figure 3.44
Global Wireless Infrastructure: Cumulative W-CDMA Contract Activity.

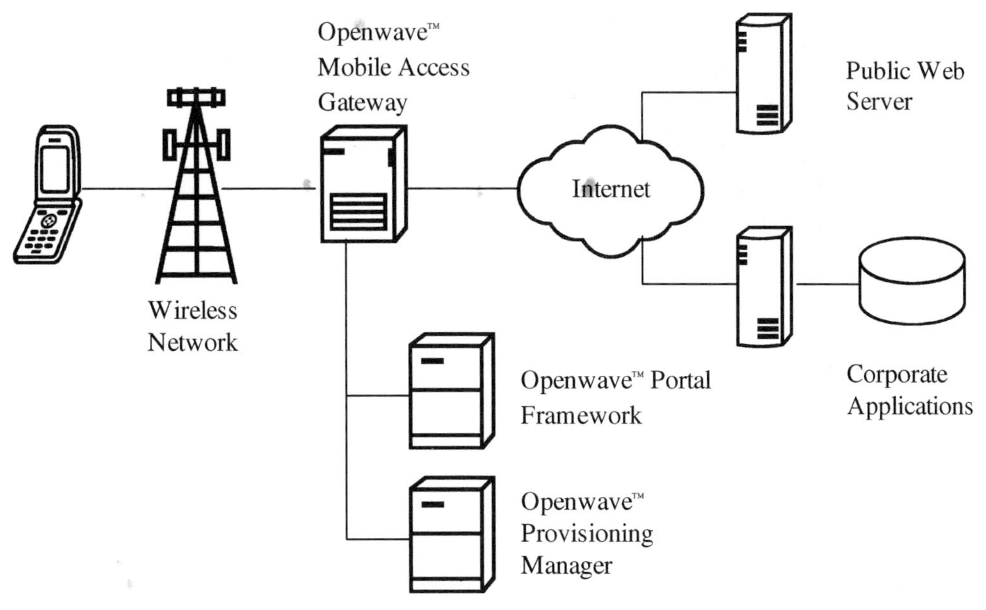

Figure 3.45
Openwave's Wireless Internet Solution.

particular the infrastructure industry, as the migration from circuit-switched to packet-switched networks occurs (see Figures 3.46 and 3.47).

Even carriers that do not want to invest in additional infrastructure, due to limited returns, may have to invest to remain competitive. This should further drive infrastructure equipment growth.

Structural Market Assessment

The wireless infrastructure market is structurally attractive and reasonable profitability should be expected for its participants (see Figure 3.48). This is probably the most attractive segment of the wireless industry in terms of market dynamics and profitability, and this should be expected to continue throughout the transition to 3G. This is due in large part to the high degree of intellectual property content in the product, the need for superior integration skills, and frequent capacity constraints. Data, 2.5G, and 3G will continue to propel strong growth in the industry for the foreseeable future.

GSM/TDMA Path to 3G

95A ⟶ 95B ⟶ MC 1X ⟶ MC 3X

Packet Data Equipment requirements	GSM CSD (Circuit Switched Data)	GPRS (General Packet Radio Service)	EDGE (Enhanced Data rates for GSM Evolution)	IMT-2000 CDMADirect Spread (CDMA DS)
Handset	No packet data capability -Single-Mode phones	New handsets GPRS-- enabled handsets will work on GPRS enabled networks and 9.6Kbps on GSM networks using CSD-Dual Mode phones	New handsets EDGE-- handsets will work at up to 384Kbps on EDGE enabled networks on GPRS enabled networks and 9.6Kbps on GSM networks using CSD-Tri-Mode phones	New handsets CDMA DS handsets will work at up to 2Mbps and only on 3G networks-Quad-Mode phones
Infrastructure	No packet data capability	New packet overlay/ backbone needed for circuit switched network	Further backbone modifications required	New infrastructure roll out with existing interconnect
Technology Platform	Current GSM/TDMA Technology	GSM TDMA platform with additional packet overlay	Modulation changes required to GSM TDMA platform	New CDMA infrastructure

CDMA Path to 3G

Packet Data Equipment requirements	95A	95B	IMT-2000 CDMAMulti-carrier 1X(MC 1X)	IMT-2000 CDMAMulti-carrier 3X(MC 3X)
Handset	Standard 95A handsets will work on all future networks: 95B, 1X and 3X at 14.4Kbps-Single-Mode phone *	Standard chipsets in 1999 95B handsets will work on 95A networks at 14.4Kbps and 95B, 1X and 3X systems at speeds up to 114 Kbps-Single-Mode phone	1X standard in chipsets in 2001 1X handsets will work on 95A networks at 14.4Kbps, 95B Networks at speeds up to 114 Kbps and 1X and 3X networks at speeds up to 307Kbps-Single-Mode phone	New handsets 3X handsets will work on 95A networks at 14.4Kbps, 95B networks at speeds up to 114Kbps and 1X networks at speeds up to 307 Kbps and 3X networks at 2Mbps-Single-Mode phone
Infrastructure	Standard	New software in BSC (Base Station Controller)	1X requires new software in backbone and new channel cards at base station	Backbone Modifications. New channel cards at base stations
Technology Platform	CDMA	CDMA	CDMA	CDMA

Source: : W. Carley and S. Buckingham (Mobile Lifestreams Limited)

Figure 3.46
IMT-2000 Evolution Path Options for CDMA and GSM.

	System Switching Architecture	Investment per Base Station (USD)	Investment to 2005 1800 MHz Operator per 50m people Base Amount to Meet Coverage Requirement	Investment to 2005 900 MHz Operator per 50m people Base Amount to Meet Coverage Requirement	Data Rate Kbps
HSCSD	*Circuit*	$4,000	$20m	$20m	14.4–57.6
GPRS	*Packet*	$20,000	$80m	$120m	115–184
EDGE	*Packet*	$70,000	$350m	$550m	184–384
IMT-2000, (UMTS)	*Packet*	$100,000	$2,500m	$3,500m	384–2000

Increasing bandwidth represents a substantial capital expenditure for network operators. This is in addition to very expensive spectrum requirements.

Source: Merrill Lynch, 2000

Figure 3.47
GSM Upgrade Costs.

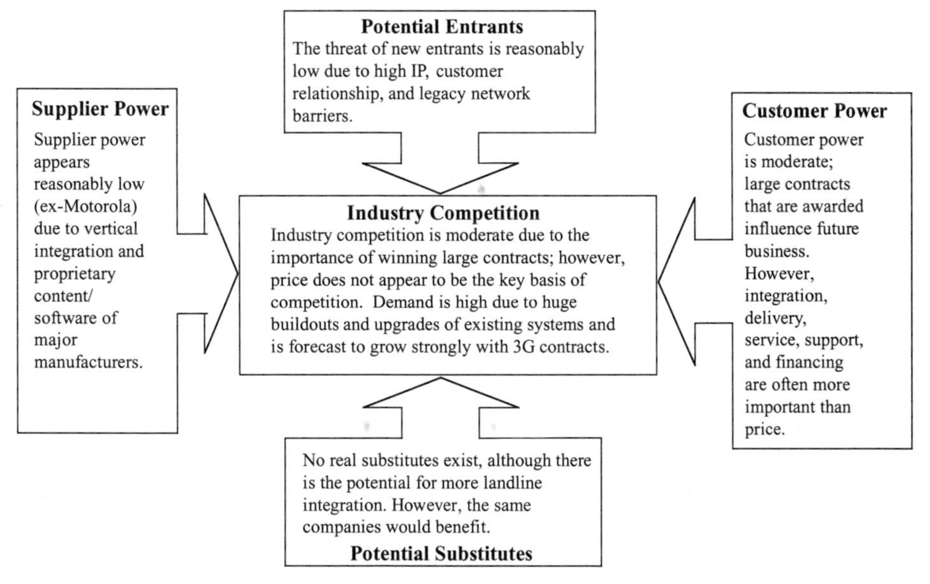

Figure 3.48
Structural Market Assessment: Infrastructure Suppliers.

Customer Power

Customer power is the issue with the greatest impact on the wireless infrastructure suppliers. They deal with large, sophisticated buyers over a long period of time to get the order. There is also a complex and competitive bidding process and orders are generally very large and influence future infrastructure decisions. However, there are many dimensions of the sale that may take precedence over price in the carrier's decision process. Some of these issues are integration with the existing wireless and landline networks, delivery, quality, reliability, durability, service, and support. In the past infrastructure suppliers have been capacity constrained due to the above forecast growth rates in mobile wireless communications. Another issue that has gained in importance is willingness to finance a portion of the transaction. This favors infrastructure suppliers with strong balance sheets, but also increases their risk.

Potential Substitutes

There are no real substitutes for wireless infrastructure. More landline integration could occur; however, it is likely that the same companies would benefit.

Supplier Power

Supplier power is reasonably low in the wireless infrastructure industry due to a high degree of vertical integration and proprietary content in the product. The exception is Motorola, which relies in part on Cisco for its switching technology.

Potential Entrants

The threat of new entrants is relatively low due to the high intellectual property content of the product and its integration into a network environment. There are also significant barriers to entry, in terms of existing relationships with carriers and integration with legacy network equipment. The carriers are very concerned about extending their existing network as much as possible, rather than completely replacing their network. This gives companies that can leverage existing carrier investments a distinct advantage.

Telecom providers are also very concerned about their quality of service and the potential impact the introduction of new equipment may have on their network. This makes them uneasy about trying new vendors in mission-critical applications and gives incumbent suppliers a large advantage.

Industry Competition

The combination of these market forces leads to a competitive marketplace, but the nature of the competition is not destructive. The main points of competition do not appear to be entirely price related, which leads to an industry that is reasonably profitable for incumbent participants. Competition appears to hinge on quality, integration skills, service, support, and the ability to provide some sales financing (particularly for new carrier entrants).

As we would expect from the preceding market analysis, the infrastructure market is very profitable, especially when compared to the profitability of other participants along the wireless communications value chain (see Figure 3.49).

Outlook

The outlook for mobile wireless infrastructure is favorable, and profitability should be expected to continue due to favorable market dynamics. The industry should grow rapidly based on increased penetration of wireless subscribers, increased minutes of use, data services, and 2.5G and 3G network builds. Carriers who do not lead in the build out of data services may well be compelled to follow quickly or risk becoming noncompetitive.

It appears that the predominant 3G technologies will be W-CDMA and CDMA2000. W-CDMA will probably be the most popular standard worldwide, but in the U.S. CDMA2000 will also be relevant. In fact, it appears that the CDMA operators have a better path to 3G than the TDMA/GSM operators.

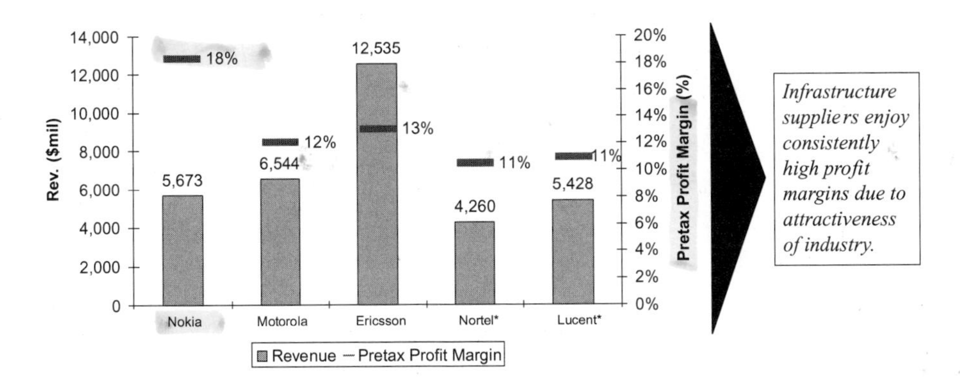

* Nortel and Lucent margins are for all products, not just wireless networking, wireless only likely to be higher.

Figure 3.49
Infrastructure Supplier Revenue and Margin Analysis.

LOCATION TECHNOLOGIES . . . .

Location technology is the ability to identify the location of a mobile terminal, and hence the subscriber, with a reasonable degree of accuracy, as it moves throughout the mobile wireless network coverage area. Location information represents the holy grail to mobile-commerce marketers. However, location technology has its roots with the FCC and its Enhanced 911 (E-911) mandate. A major issue with location services will be individual privacy rights and third-party access to location information.

E-911

The FCC has determined that it is critical to accurately identify the location of a caller when a 911 call is placed, so that emergency services can be properly dispatched. This is not a problem when an emergency call is placed from a landline phone since location and identity information are known, based on the phone number of the caller. However, 911 calls from mobile wireless phones provide only the phone number and cursory location information. The location information is limited to the cell where the call originates. This function is available only in regions where the capability has been activated.

Accurately ascertaining location has become a very important issue because a large percentage (30+% according to Robinson-Humphrey, about 50% according to Lehman Brothers) of 911 calls are placed from wireless handsets (Bensche 2000, 5). When a call is placed from a mobile wireless phone, the caller often cannot accurately specify location. For example, if one witnesses an accident on the highway, it is often difficult to specify precise location information. This becomes more of an issue if callers are incapacitated or do not know their location.

Since April 1, 1998, Phase I of E-911, the FCC has required wireless carriers to transmit all wireless 911 calls to emergency assistance providers operating Public Safety Answering Points (PSAPs). The wireless carriers are required to provide the telephone number and cell site information to the PSAP. This allows the PSAP to call the individual back if the call is disconnected, and gives a rough indication of the caller's location. Location by cell site is not precise enough to adequately dispatch emergency services, but does provide neighborhood/town scale location information.

According to the National Emergency Number Association, in November 1999 only 7% of PSAPs were capable of processing Phase 1 information. As of June 2000, only 30% of PSAPs had requested Phase 1 emergency services. The reasons cited for the slower than expected roll-out were limited funding and implementation delays at the PSAPs. AT&T was the only carrier in Phase 1 compliance on April 1, 1998, and as of summer 2000 *Wireless Week* reported only 5% were in compliance (Bensche 2000, 7). Carriers cited lack of requests by PSAPs as the reason.

Funding for the PSAPs is usually collected by the carriers through a 911 surcharge. This surcharge ranges from state to state, but is usually $0.70–$1.00 per month per customer. The charge is forwarded by the carrier to the state tax collector who turns it over to the PSAPs for cost recovery.

Phase II of the FCC's E-911 rules requires wireless carriers to have Automatic Location Identification (ALI) capability in order to pinpoint caller location more precisely. Phase II is supposed to be operational by October 1, 2001. The FCC does not require a specific technology to be used to meet its mandate, but does require adherence to performance metrics. Carriers may employ a location technology that is either network based or handset based. However, the technology employed must meet the following standards for accuracy and reliability. For network-based technology, location accuracy must be within 100 meters 67% of the time and within 300 meters 95% of the time. For handset-based technology, location must be provided within 50 meters 67% of the time and 150 meters 95% of the time.

	67% of Time	**95% of Time**
Handset Solution	50 meters	150 meters
Network Solution	100 meters	300 meters

These are the current standards and dates. However, both may slip if carriers are unable to achieve these standards after demonstrating good-faith efforts to comply.

Location Methods

The main classes of technology for ascertaining location can be categorized as either network-based or handset-based solutions.

Cell of Origin (COO)—Network Based

Cell of Origin (COO) is the system currently used to comply with Phase I E-911 requirements. This technology tells what cell a caller is presently occupying, but offers no greater resolution. Cell location can generally be ascertained in about three seconds. The accuracy of COO is determined by the teledensity of the area, with accuracy being proportional to the number of cell sites or the size of the cell. This solution requires no alteration to the network or to the handsets, but is insufficient for emergency services.

Angle of Arrival (AOA)—Network Based

Angle of Arrival (AOA) uses an array of large antennas at the cell site to measure the angle of the incoming control signal from the handset. A minimum of two cell sites are required to determine location, and no handset modifications are needed. However, this solution requires substantial capital expenditures by the carrier, particularly in rural areas.

AOA accuracy is negatively impacted by line-of-sight obstacles and distance from the base station, and may not function well in an urban environment. AT&T has tested AOA, but fears that it will be difficult to get permission to erect large, unsightly antennas, and feels that both property owners and zoning boards will reject their use.

AOA alone is unlikely to be a suitable solution. However, AOA has shown promise as a hybrid solution with TDOA (see below). A combination AOA+TDOA solution is a stronger offering than either separately. Hybrid AOA+TDOA solutions are supplied by Grayson Wireless, True Position, Sigma One, and Radix Technologies.

Time Distance of Arrival (TDOA)—Network Based

Time Distance of Arrival (TDOA) uses at least three base stations to measure and compare the arrival time of the control signal from a mobile handset in order to calculate location. To accurately determine location, strict synchronization of the base stations is required. Synchronization is such a critical issue that an atomic clock is used in each base station. This solution may be attractive to CDMA networks, which are already synchronized, versus GSM networks, which may be asynchronous.

TDOA requires line of sight to determine location. This can present problems in rural areas where three cell sites cannot be accessed simultaneously, and urban canyons where multipath reflection can be a problem. TDOA is also less accurate than E-OTD, Cell-ID, or GPS, and can take up to 10 seconds to determine location.

TDOA antennas are less expensive and easier to deploy than AOA antennas, and TDOA does not require any handset modifications. However, TDOA is still regarded as a fairly expensive solution.

Suppliers of TDOA systems include Cell-Loc, True-Position, and Grayson Wireless.

Multipath Fingerprinting (MF)— Network Based

Multipath Fingerprinting (MF) locates the caller by matching the received radiowave to a reference radiowave in the system database. The reference radiowave, or fingerprint, stored in the database takes into account the wave reflections generated by making a call from a specific location. This fingerprint is then matched to the actual call and location is estimated. Due to the dynamic nature of the environment, maintaining an up-to-date database is critical.

Only one cell site is needed to determine location and no handset modifications are required. MF is regarded as the most accurate and fastest network-based system, and can usually fix location within 86–100 meters in one to two seconds. This system is regarded as fairly inexpensive to deploy.

U.S. Wireless is the primary supplier of this technology.

Enhanced Cell ID (E-CID)—Network Based

Enhanced Cell ID (E-CID) is a software-based solution that determines location by comparing the list, or table, of cell sites available to the handset. Once the available cell sites are known (this is constantly updated), location can be calculated based on the intersections of the overlapping cells.

This system works best in areas with many cell sites, and location can be determined within about 100 meters (250 meters rural). A key advantage of E-CID is that line of sight is not required. Currently, this system works only with GSM networks. E-CID requires slight modification to the SIM card in the handset and a proprietary network server. Therefore, it is regarded as a relatively low cost and nondisruptive solution for GSM operators.

The largest suppliers of E-CID systems are Cellpoint and CT Motion.

Enhanced Observed Time Difference (E-OTD)—Hybrid

Enhanced Observed Time Difference (E-OTD) operates under the same principles as TDOA (measuring the time it takes to receive a signal), but in the reverse. The signals are received from at least three base stations, whose locations are known, and location is calculated by the handset.

E-OTD uses the existing capabilities of the GSM protocol and is relatively straightforward to apply to these networks. E-OTD is a more costly and complex solution to deploy because of handset software upgrades and location measurement units (reference beacons), but it yields much better location information. Location measurement units are distributed throughout the network, with about one unit for every four cell sites. Again time keeping is of the utmost importance, and system time is usually kept by an atomic clock.

E-OTD can usually provide location information accurate to 50–125 meters within five seconds. However, this system can be susceptible to distortion in urban areas.

Cambridge Positioning Systems has developed its Cursor product based on E-OTD.

Global Positioning System (GPS)—Handset-Based Solution

GPS is a worldwide radio-navigation system formed of 24 satellites and ground stations sponsored by the U.S. Department of Defense. The system measures the longitude, latitude, and elevation of the receiver. Again, triangulation is used to determine location. This is accomplished by measuring the time it takes to communicate with three satellites and then calculating position. A fourth measurement is taken to ensure that the timing of the pseudorandom codes is synchronized. Again, time is critical in the calculation of location, and an atomic clock is used.

Until May 1, 2000, the U.S. government used Selective Availability (SA) to disrupt the GPS system and degrade its accuracy to protect against hostile forces using GPS for weapons systems. This degraded accuracy to about 100 meters. With SA discontinued, GPS is accurate to 10–20 meters. Differential GPS can correct the various inaccuracies in the GPS system, including SA, and it is accurate to about 2 meters.

GPS requires line of sight in order to calculate location. This means that location cannot be determined if the user is inside a building, in an urban canyon, or under a heavy canopy of trees. GPS takes the most time to determine location, requiring 10–60 seconds. This amount of time is needed to communicate with the satellites and to perform the complex calculation.

GPS is enabled through a chipset in the handset and a second antenna. This increases the cost (estimates range widely: $25–400), the size, and the power consumption of the handset. For GPS to work throughout a network every handset would need to be replaced. This is a very expensive undertaking, and would likely require subsidization by the carrier. Factors working in favor of the network operator include the normal replacement cycle for handsets of about every 18–24 months, high industry churn levels, and historically declining prices for chipsets.

Assisted Global Positioning System (A-GPS)—Hybrid

Assisted GPS helps to overcome some of the drawbacks of pure GPS such as cost, power consumption, speed to determine location, and the line-of-sight requirement, by shifting much of the processing burden from the handset to the network. Additionally, the network keeps track of location so that when satellites are obstructed, a good estimate of location can be obtained based on the last reading.

A-GPS is less costly (about $20) from a handset perspective, but requires additional investment in the network. Location can usually be ascertained in about 5 seconds, and A-GPS accuracy is considered the highest.

SnapTrak, a division of Qualcomm, is one of the leaders in this field. SiRF, IDC, and Cellpoint are also active in this technology.

See Figure 3.50 for a comparison of the location technologies discussed here.

Carrier Location Compliance

The wireless carriers seem to be well behind schedule in implementing Phase II of E-911. Their reports on Phase II compliance, due October 2000, were characterized by the FCC as very disappointing. When this is considered against the backdrop of Phase I compliance, which has been unsatisfactory, particularly from the PSAP request perspective, it may be likely that the mandate will ultimately be delayed past October 2001.

Service Bureau Concept

Several firms are considering a service bureau approach. Under this approach, a firm would build a location-identifying infrastructure and then sell the service to a number of carriers. This would lower the deployment cost and reduce the technological risk

	GPS	A-GPS	E-CID	E-OTD	AOA	TDOA	AOA+ TDOA	Finger-print
Accuracy	★★★	★★★★	★★★	★★	★	★★	★★	★★★
Carrier Costs	★★★★	★★★	★★★	★★★	★★★★	★	★★	★★
Subscriber Costs	★	★★	★★★★	★★	★★★★	★★★★	★★★★	★★★★
Potential Churn	★	★★	★★★★	★★	★★★★	★★★★	★★★★	★★★★
Power Use	★	★★	★★	★★	★★★★	★★★★	★★★★	★★★★
Time to Locate	★	★★★	★★★	★★★	★★★	★★★	★★★	★★★★

Source: Lehman Brothers, 2000

Figure 3.50
Comparison of Location Technologies.

for the carrier. This information could potentially be sold to m-commerce companies as well.

U.S. Wireless (Multipath Fingerprinting) and TimesThree (TDOA), a division of Cell-Loc, are two companies offering the service bureau concept.

Location-Based Applications

As a result of location-identifying technology, a myriad of revenue-generating location-based services can be made available. Some of the major classes of applications can be tracking, guiding, and notification. These services become even more compelling in a packet-based environment, where information can be pushed to the recipient, instead of having to be speculatively retrieved.

Tracking applications would be valuable in locating and tracking assets like fleets. Other potential uses include ascertaining the location of children, employees, pets, or even criminals.

Guiding applications could assist customers in finding specific locations and addresses. These services could also be used to find the nearest Chinese restaurant, ATM, or shoe repair. This application is considered to be a very attractive advertising and revenue-generating application.

Notification applications can be used to alert subscribers regarding stock trades, traffic situations, flight delays, and email. This application is particularly useful when enabled by a packet-switched network where information can be proactively pushed to the subscriber in a timely fashion.

Market Size and Forecast

Location-based services are expected to grow very rapidly and generate substantial revenues (see Figures 3.51 and 3.52). One of the key concerns with making location information available is privacy. People are concerned that the government, employers, or businesses will have too much ability to track their movements. This may lead to a preference for handset-based solutions if the locating apparatus can be turned off. This will also be an issue for m-commerce, as users may not appreciate having unsolicited advertisements sent to their handsets. This is particularly true if there is some charge to the user for receiving the communication.

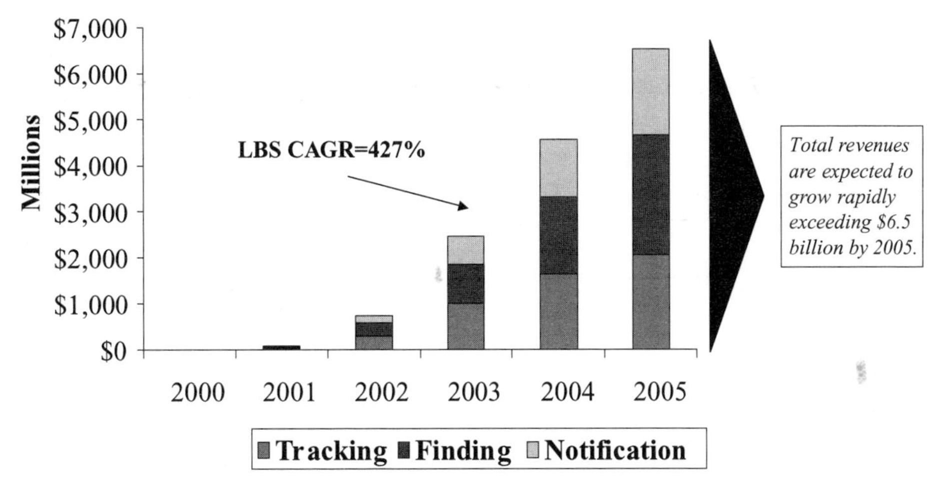

Source: Strategy Analytics, Inc.: a Boston-based company (2000b)

Figure 3.51
Location-Based Services Revenues by Application.

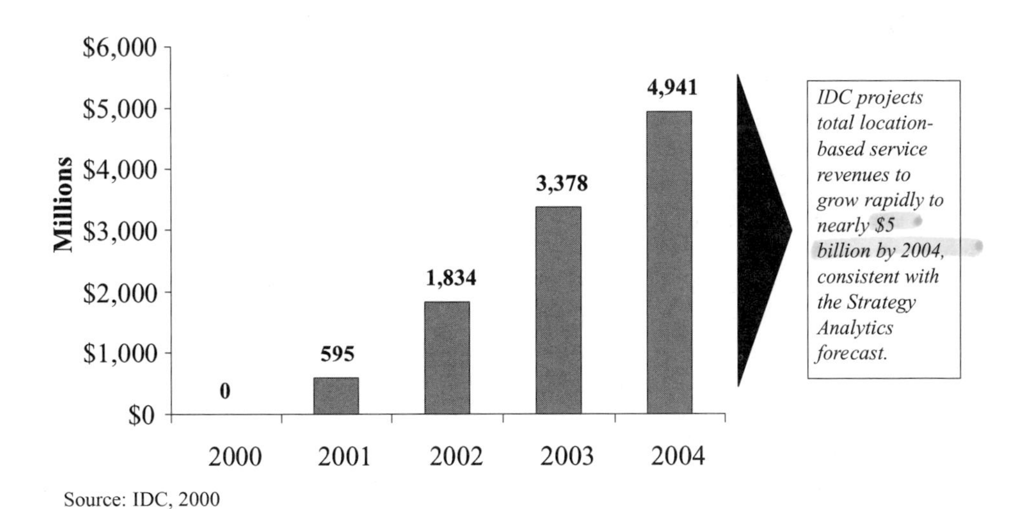

Source: IDC, 2000

Figure 3.52
Location-Based Services Total Revenues.

Outlook

Clearly, location-based services offer a tremendous opportunity for the mobile wireless communications industry. First of all is the increase in safety through location-aware 911 access, which was the original driving force behind implementing the technology. Additionally, the marketing and service opportunities are likely to prove enormous.

From a marketing perspective, it is tremendously valuable to know the identity of an individual, the location of an individual, and what the individual is most interested in at that particular moment. Additionally, the ability to proactively push relevant, timely, and location-specific information to a subscriber is extremely valuable.

Conversely, it seems less clear that pushing advertisements or coupons to individuals will be successful. Many industry pundits believe that people will respond favorably to a situation where they are in the vicinity of a store and they are alerted to a $10-off coupon. However, it is easy to imagine that consumers would react negatively to this situation and view it as irritating and intrusive.

Since little progress has been made in complying with E-911, no technology has achieved critical mass and it is too early to predict which technology will achieve dominance. It is likely that, in the U.S., the decision will be made based on a carrier-by-carrier analysis which will lead to several technologies being adopted. This will complicate location determination when a subscriber is roaming if the handset is not compatible with the technology in use in the particular area.

In total, the addition of location capability to the wireless marketplace will be invaluable both to consumers and businesses. However, it will require responsible management to allay privacy concerns and many technical and financial barriers will need to be overcome.

4 Carriers

In this chapter...

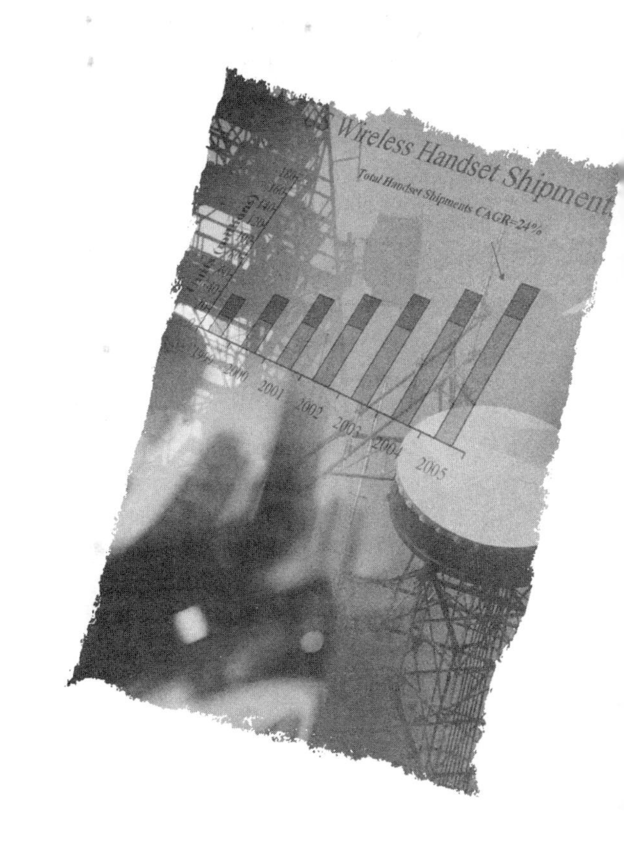

Network service providers, or carriers, are firms that have secured spectrum and built a mobile wireless communications network. The largest mobile wireless communications market in terms of revenue is cellular telephony. Voice is the killer app of mobile wireless communications, but data is expected to gain rapidly in importance as high-speed, packet-switched networks become more prevalent. Data/paging carriers are expected to face difficult market conditions as a result of intense competition from the cellular carriers.

MOBILE WIRELESS VOICE . . . .

A carrier, in mobile wireless communications, is a company that provides wireless telephony and data services to its customers over its network. Carriers are the main protagonists in the wireless arena, and the vast majority of industry revenues accrue to them. They own the relationship with the customer, determine which transmission technology to use, select the handsets that are authorized to operate on their network, and handle billing for services. The major wireless carriers in the U.S. market are Verizon Wireless, Cingular Wireless (a combination of SBC and BellSouth wireless operations), AT&T Wireless, Sprint PCS, VoiceStream, and Nextel.

Market Size and Forecast

U.S. mobile wireless communications service is an enormous market that has experienced tremendous growth over the past decade, as the cellular phone has evolved from an elite status symbol to a mass-market consumer item. In 2000, the U.S. mobile wireless market generated more than $40 billion in service revenues, with subscribership in excess of 105 million and market penetration greater than 35% (see Figures 4.1–4.3). Strong growth is expected to continue for the foreseeable future. It is predicted that by 2005 there could be as many as 200 million subscribers (80% market penetration) in the U.S., generating service revenues in excess of $100 billion.

The mass-market acceptance, and amazing level of growth over such a large base, has led to the palpable excitement surrounding the mobile wireless communications industry. The future is even more exciting, from a growth perspective, when data and 3G services are factored into the equation. However, the current structure of the industry is fiercely competitive with falling prices and low returns.

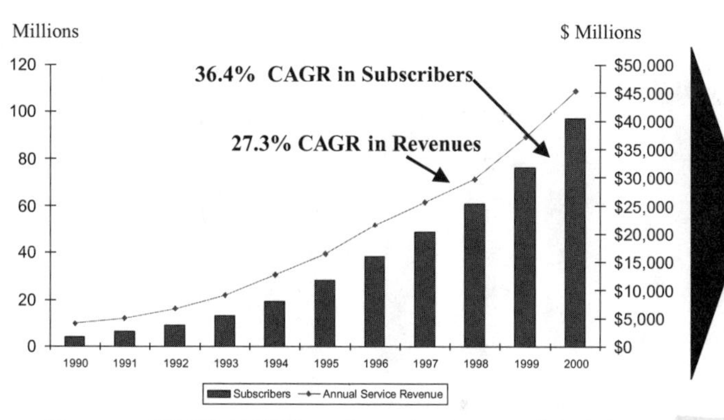

The U.S. wireless telephony industry has grown subscribership at more than 36% and revenues at more than 27% per annum for the last 10 years. A phenomenal rate of growth for an industry this large.

There were 105,760,056 U.S. wireless subscribers as of November 17, 2000

Source: CTIA, 2000 (Measured June to June)

Figure 4.1
U.S. Mobile Wireless Telephone Subscribers and Revenues 1990–2000.

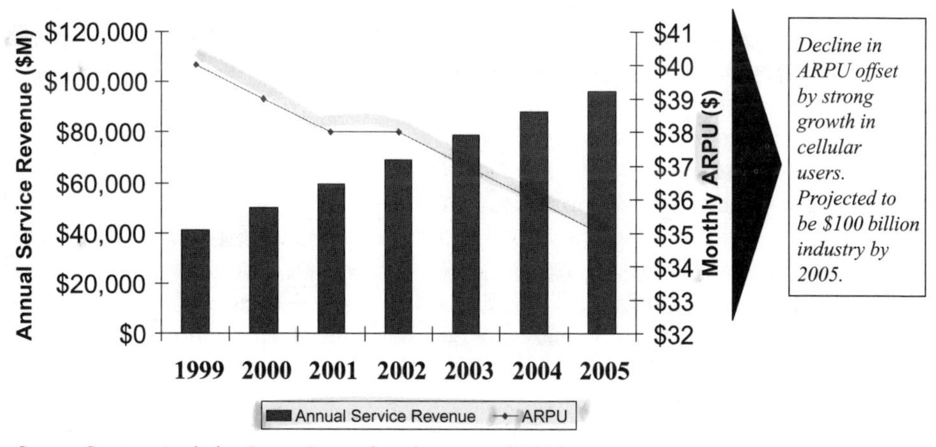

Source: Strategy Analytics, Inc.: a Boston-based company (2000d)

Figure 4.2
U.S. Mobile Wireless Service Revenue and Monthly Average Revenue per Unit (ARPU).

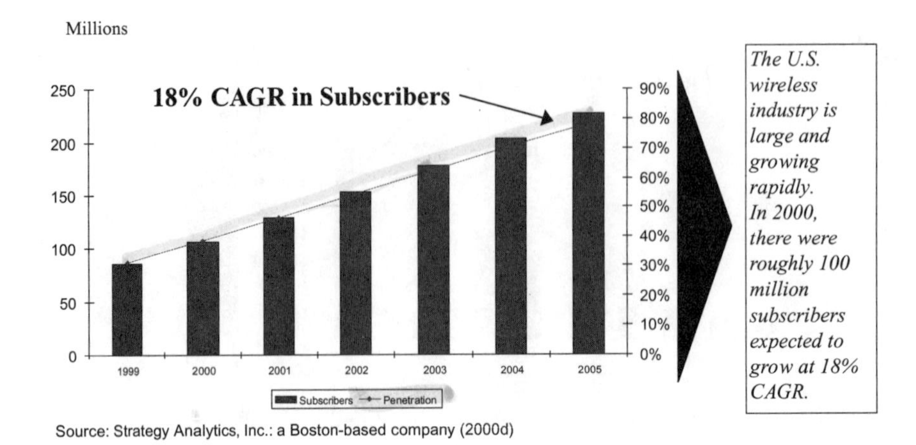

Source: Strategy Analytics, Inc.: a Boston-based company (2000d)

Figure 4.3
U.S. Mobile Wireless Telephone Subscribers and Penetration.

Major Carriers

The U.S. mobile wireless telephony market is becoming increasingly characterized by the megacarrier. These are carriers that offer virtually nationwide coverage, no incremental charge for long distance, and frequently no roaming charges. Roaming is when subscribers want to use mobile wireless service away from their home area (as defined by their calling plan) or want to access service through another carrier's network where their primary carrier doesn't have a presence. The largest carriers are continuing to acquire smaller operators as the industry consolidates (see Figure 4.4), due to the need for scale in operations and marketing. In 2000, the largest wireless carrier was Verizon Wireless. Verizon Wireless represents the combination of GTE Wireless, Bell Atlantic Wireless, and Vodafone (formerly Airtouch). As of the third quarter of 2000, they had more than 26 million subscribers representing about a quarter of all subscribers (see Figures 4.5 and 4.6). The megacarriers, combined, control over 70% of all subscribers.

Performance Drivers

The key drivers of operating performance in the mobile wireless communications industry are number of subscribers, acquisition cost, customer churn, and Average Revenue per Unit (ARPU). These four issues drive the operating results of the business and are within the control of the carrier.

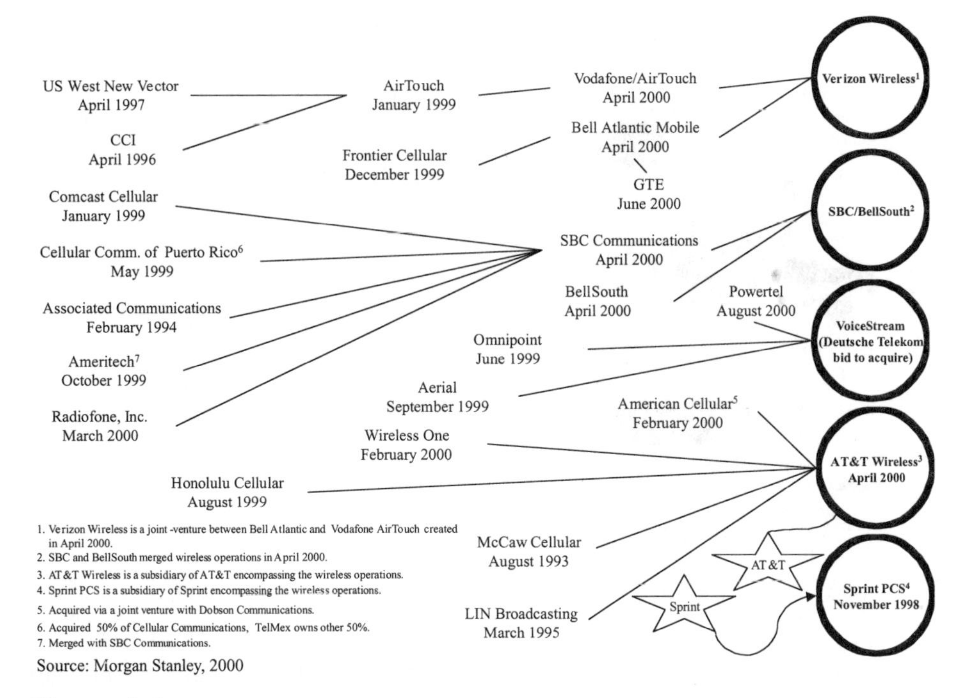

Figure 4.4
U.S. Mobile Wireless Service Provider Consolidation.

Service Provider	Subscribers 3Q00 (Millions)	Covered POPs (Millions)	Technology
Verizon (Bell Atlantic & Vodafone)	26.2	210.0	CDMA
Cingular (SBC/BLS)	19.2	175.0	CDMA, TDMA, GSM
AT&T Wireless	12.6	121.0	TDMA
Sprint PCS	8.4	176.0	CDMA
Nextel	6.2	175.0	IDEN
Voicestream	3.9	94.0	GSM

A number of nationwide providers are emerging to dominate the wireless market in the U.S. They offer digital service and one-rate national plans with no long distance charges and in many cases no roaming charges.

1. SBC converting network to TDMA

Figure 4.5
Mobile Wireless Service Provider Summary.

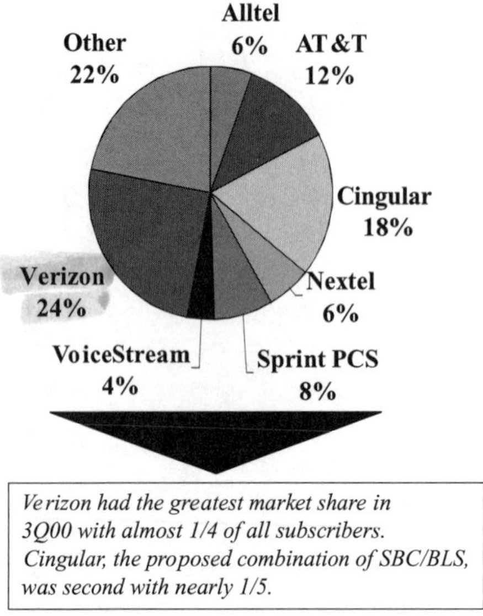

Source: CS First Boston and CTIA

Figure 4.6
Mobile Wireless Communications Subscriber Market Share Third Quarter 2000.

The economics of provisioning a wireless network are similar to that of most other telecommunications networks. They entail high fixed/sunk costs (spectrum license and infrastructure) and relatively low operating costs. Telecommunications networks have typically been regulated monopolies due to the nature of this investment and the significant potential for hypercompetition. However, in mobile wireless communications there are a large number of competitors in every major market. This situation is enforced by the FCC through spectrum caps. As of January 1, 2001, FCC spectrum caps ensured that there were at least five competitors in metropolitan markets and at least four competitors in rural markets.

In a market with several participants, this economic structure can be expected to lead to intense competition and rapid consolidation. This is what is happening in the industry and should be expected to continue. Historically, the industry has grown very rapidly; however, prices are falling and wireless carriers, with the exception of AT&T Wireless and a few rural operators, are not producing any earnings (see Figure 4.7).

Customer Acquisition Cost

Customer acquisition costs (see Figure 4.8) are one of the key drivers of operating results. Customer acquisition costs are very high and average $350–400 per subscriber.

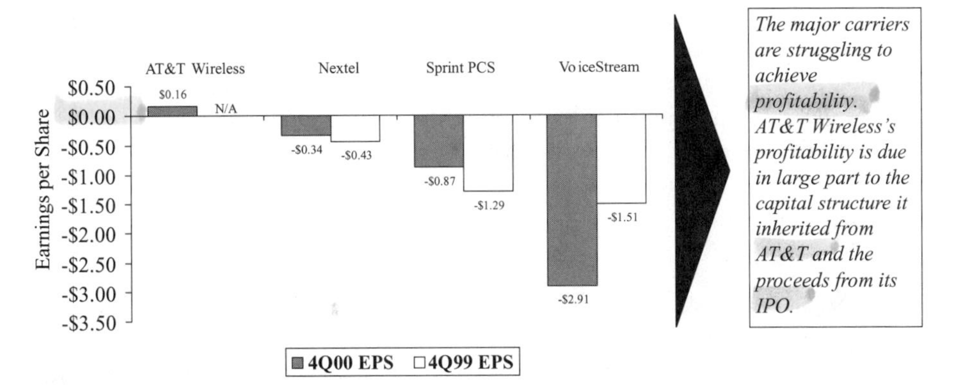

Figure 4.7
Major Carriers' Profitability Levels, Fourth Quarter 2000 vs. Fourth Quarter 1999.

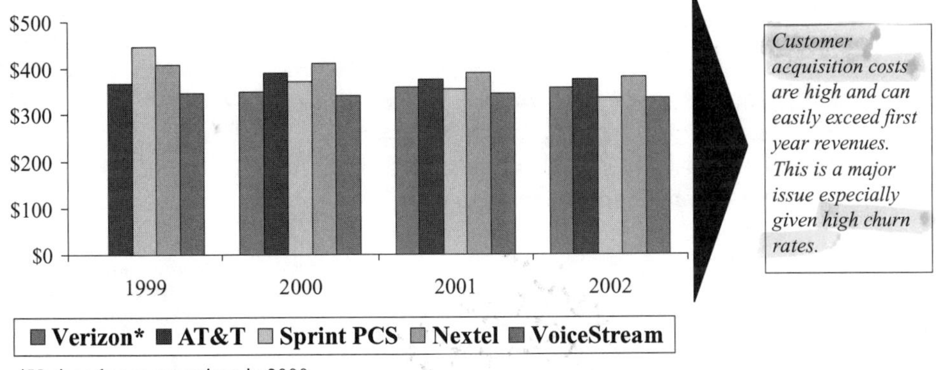

Figure 4.8
Customer Acquisition Cost.

The cost of acquisition is even more substantial when it is measured against ARPU. With ARPUs in the $40–45 per month range, it takes about nine months of revenues to recoup the acquisition cost. The time it takes to recoup customer acquisition costs may be longer in the future, as incrementally less intensive users are being added to the mix and ARPU/pricing is declining faster than acquisition costs.

The major components of acquisition cost are equipment subsidies, sales commissions (at company store or reseller), and marketing costs. On top of acquisition costs, network capital, spectrum, and operating costs need to be recouped before a profit can be earned.

Customer acquisition costs are usually quoted on a gross customer-added basis rather than a net customer-added basis. Therefore, due to the high levels of industry churn, it is substantially more costly to add a net new subscriber than what is typically quoted. This highlights the importance of another key driver of operating results in the wireless carrier business—churn.

Churn

Churn is when wireless service is disconnected. This can happen at the request of the customer, who terminates or switches carriers, or at the request of the carrier, in the event of nonpayment for services. The level of churn in the wireless industry is truly astronomical. The industry typically experiences churn in the 2–3% per month range (see Figure 4.9). Churn figures are generally quoted on a monthly basis; when annualized, churn rates are in the 30–40% range (see Figure 4.10). This is a huge issue for carriers and, with some relatively conservative assumptions, could add up to a $20 billion per year problem for the industry.

Given the high cost of customer acquisition, it is even more critical to minimize the loss of existing customers through churn. There are very few industries where companies are able to operate profitably when they have to replace 30–40% of their customers every year just to maintain their subscribership. High customer mobility also enables new or emboldened competitors to make meaningful inroads into the market through price or service differentiation.

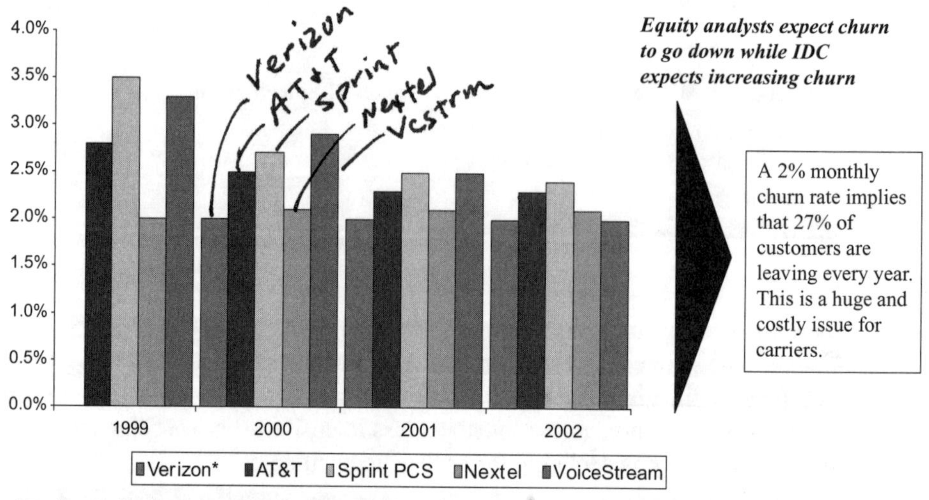

*Verizon began operations in 2000
Source: Company reports and analyst estimates

Figure 4.9
Monthly Churn Rates for Service Providers.

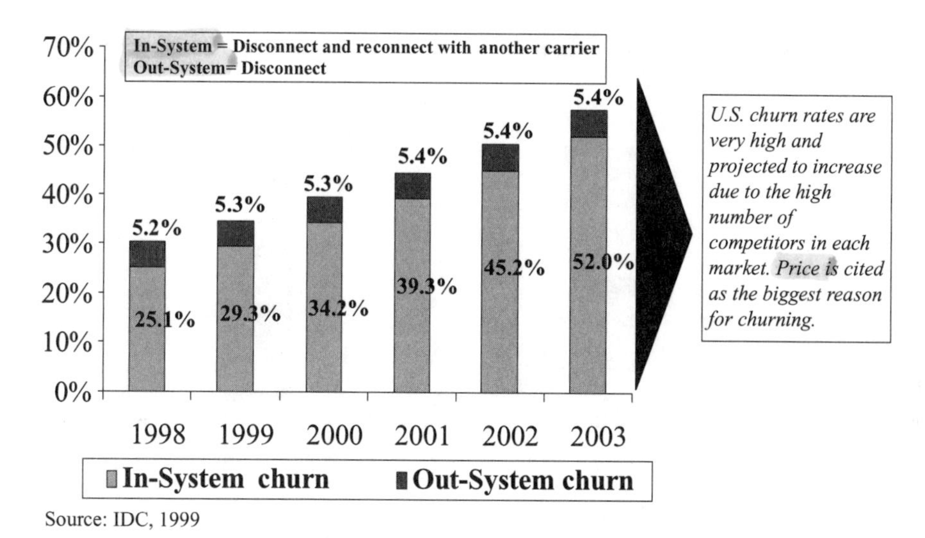

Source: IDC, 1999

Figure 4.10
U.S. Churn Rates (Annual).

There are two types of churn. In-system churn, where someone disconnects and reconnects service with another carrier, and out-system churn, where someone disconnects service and doesn't reconnect. Out-system churn includes people who voluntarily cease service, move out of the area (may reconnect at new location), or are disconnected by the carrier.

In comparing churn forecasts, it is instructive to note that equity analysts forecast decreasing levels of churn, but IDC forecasts increasing levels of churn. For reference, industry churn has increased in 2000. One of the potential reasons for the differing forecasts is that IDC does not have to develop company valuations based on operating results. This may provide less reason to be optimistic about customer retention. However, given the level and cost of churn, one would expect the carriers to develop more effective customer retention strategies over time, including acquiring competitors.

By overlaying forecast churn rates and forecasts of subscribership, we can estimate the total number of subscribers churning. By 2003, using IDC estimates, more than 60 million subscribers will be voluntarily churning within the system (see Figure 4.11). The magnitude of this problem is enormous. Over the long term, very few industries are able to operate profitably with this level of customer mobility, especially given the high cost of customer acquisition.

One of the more interesting dynamics of churn, is that people are not disconnecting service because they do not like or value the product (see Figures 4.12 and 4.13). In fact, quite the opposite, many customers find wireless services to be a good value and they are generally satisfied with their current supplier. This is especially true when compared with other communications suppliers, such as their local phone

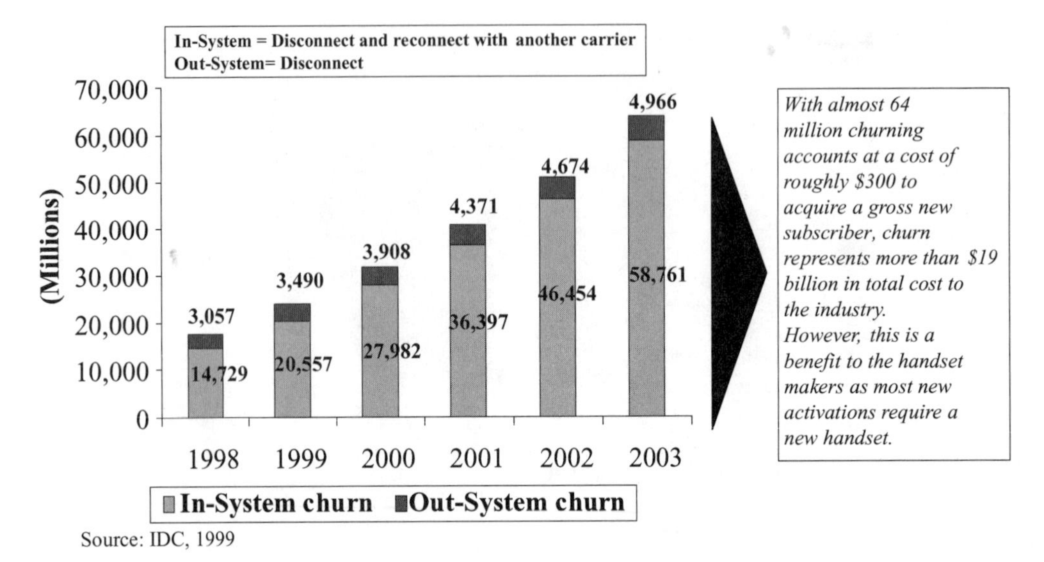

Figure 4.11
U.S. Churn Number of Subscribers (Annual).

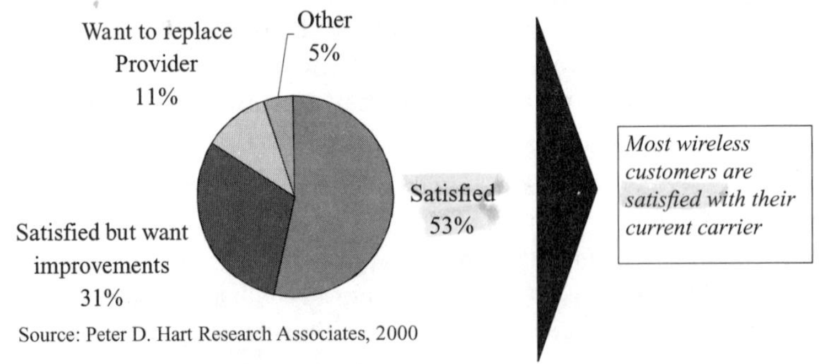

Figure 4.12
Satisfaction with Wireless Carrier.

or cable company. In fairness, however, local monopolies are rarely considered to be good values.

Given the high levels of satisfaction and the perceived reasonably high value, it is important to understand why such a large number of customers churn. The primary reason cited for churning, as reported by IDC, is a better offer from a competitor (41%, as shown in Figure 4.14). This does not bode well for the financial returns of the industry and supports the notion that wireless service is viewed as a commodity and that competition is based primarily on price. This is reinforced by the churn forecast that shows that out-system churn is only a small portion of overall churn.

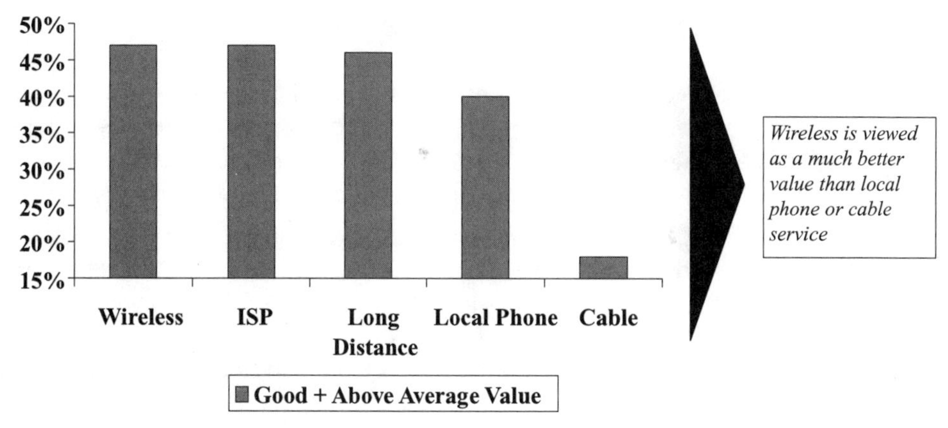

Figure 4.13
Customer Perception of Value of Communications Services.

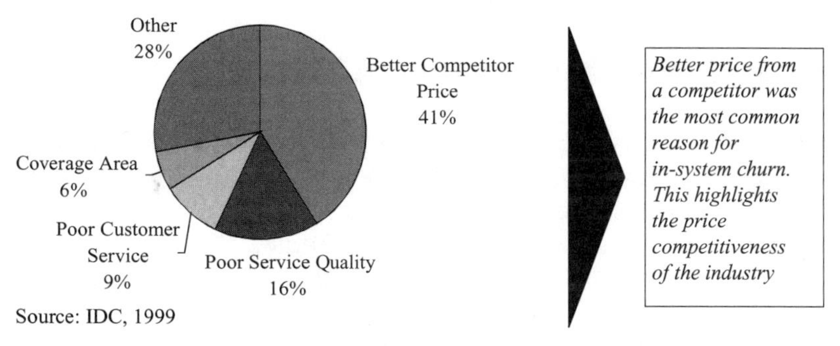

Figure 4.14
Primary Reasons for In-System Churn.

Another potential reason for high levels of churn, one that is not often mentioned, is the desire of customers to get a new and improved handset. This is particularly true as people transition from analog to digital service. In 2000, many subscribers still had analog service. Some are in areas that haven't upgraded to digital service, mostly rural, and others are "glove compartment" or security users who do not routinely use their phones. However, in the near term, there are still a large number of active users who will be coming off analog contracts and will opt for digital service. Handset-motivated churn should be expected to continue as more data services are rolled out and as the industry transitions to 2.5G and 3G services, which require new handsets to enjoy their enhanced features.

Use of Contracts to Reduce Churn

Due to the exceedingly high cost of acquisition and churn, carriers have attempted to improve retention by locking customers in through contracts (see Figure 4.15). The most typical contract term is 12 months. The 12-month contract allows the carrier to recoup its acquisition cost, but does not provide much opportunity for profit.

However, as a retention tool, contracts do not appear to be particularly effective. IDC found that the presence of a contract did not significantly reduce churn. It appears that voluntary churners simply switched service at the end of the contract period as they examined the market to see if there was a more attractive service option. It also appears that the market is unwilling to sign up for contracts longer than 12 months, especially given the speed of change in the industry and the falling price environment. This situation presents quite a quandary for the carrier with no readily apparent answer. This is one of the primary reasons carriers are interested in developing data services. With data, carriers hope that they can create personalized and "sticky" services that will help them to improve customer retention.

Average Revenue per Unit (ARPU)

Monthly mobile wireless bills have been trending downward for the past decade as the cost per minute of service has fallen steadily (see Figure 4.16). This trend may be moderating as customers are picking up large bucket-of-minute plans with no incremental charge for long distance (see Figure 4.17). In 2000, according to CTIA, the average bill actually upticked from $40 per month to $45, after falling steadily from

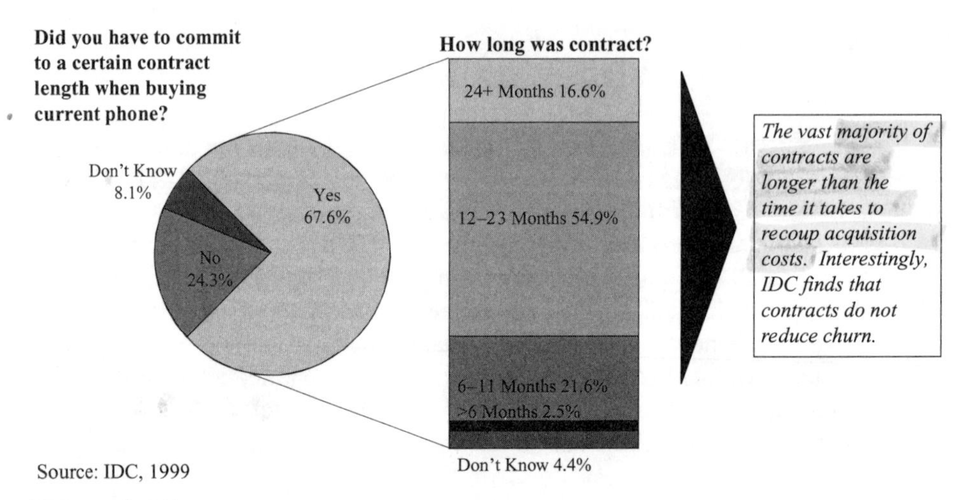

Figure 4.15
Presence and Length of Mobile Wireless Contracts.

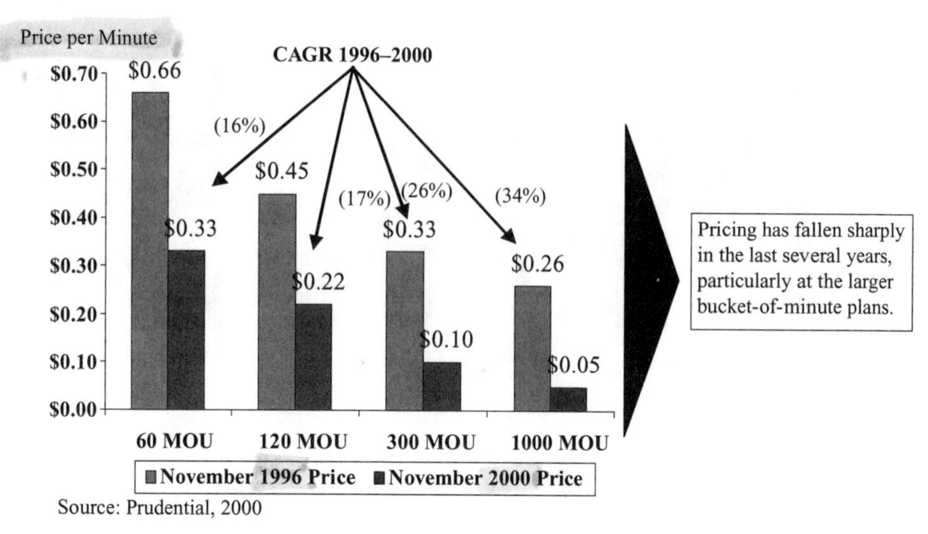

Figure 4.16
Digital Pricing per Minute of Use by Plan 1996–2000.

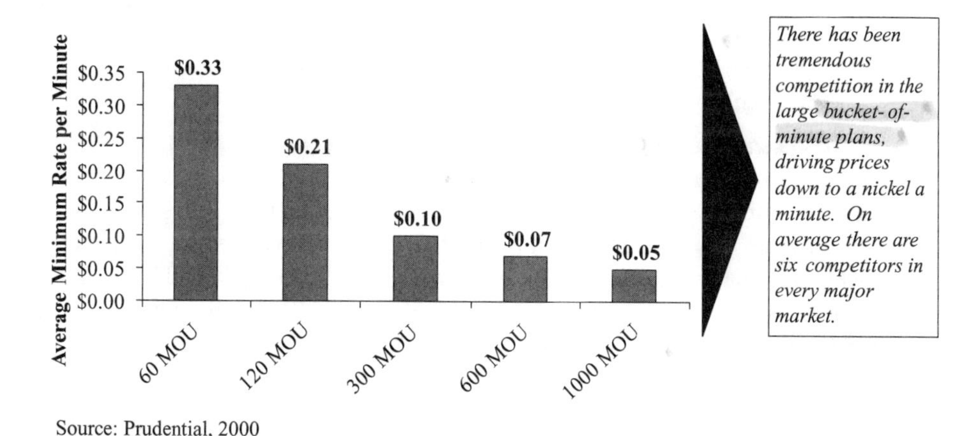

Figure 4.17
Digital Service Rates per Minute by Plan November 2000.

$84 per month since 1990 (see Figure 4.18). Despite this declining price environment, subscriber growth has been exceptional, and overall industry revenues have skyrocketed over the last decade.

The fact that revenues of the industry have gone up much faster than prices have declined over the last 10 years indicates a positive price elasticity for mobile wireless communications services, where for every 1% decline in price there has been more than a 1% increase in revenues. This feature of the market would tend to encourage

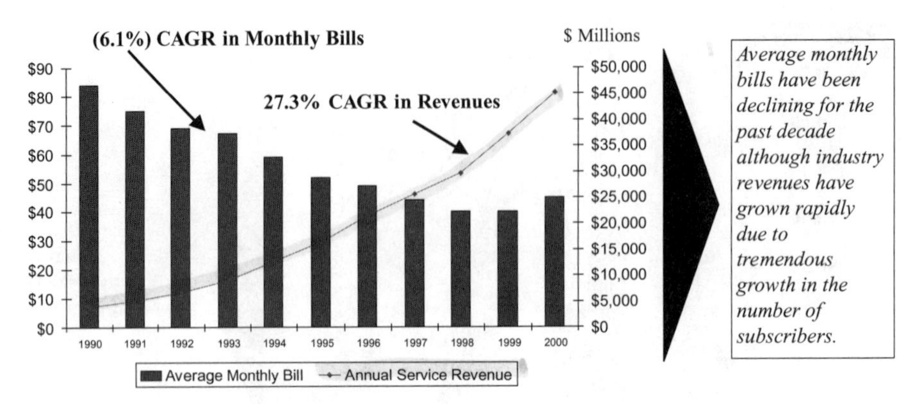

Figure 4.18
U.S. Mobile Wireless Telephone Average Monthly Bill and Service Revenues 1990–2000.

industry participants, especially second-tier firms, to cut prices in order to stimulate demand. Conversely, if firms were to attempt to raise prices they would actually reduce total revenues. This was demonstrated in dramatic fashion in the paging industry in 1999, when PageNet, the number one paging firm at the time, attempted to raise prices. They effectively raised prices by 1.2% but their revenues declined by more than 5%.

Monthly ARPU is projected to continue its downward trend as prices decline and less intensive users are added to the mix. This may be moderated somewhat by large bucket-of-minute plans, no incremental cost for long distance calls, and the advent of data services. In spite of declining ARPU, industry revenues are projected to grow briskly as a large number of new subscribers are added. By 2005, the U.S. mobile wireless communications industry could well exceed $100 billion in service revenues (see Figure 4.19).

Minutes of Use (MOU)

Minutes of use (MOU) are projected to continue to rise sharply due to lower prices, large bucket-of-minute plans, and new data services (see Figure 4.20). Minutes of use are driven by both number of calls placed and by the length of each call. Numbers of calls placed have been on the rise and the average call length has been trending upwards over the past few years. The average call length for 2000, as reported by CTIA, was 2½ minutes (see Figure 4.21).

The fact that MOUs are projected to increase in the face of declining ARPUs continues to emphasize the expectation in the industry that prices will fall.

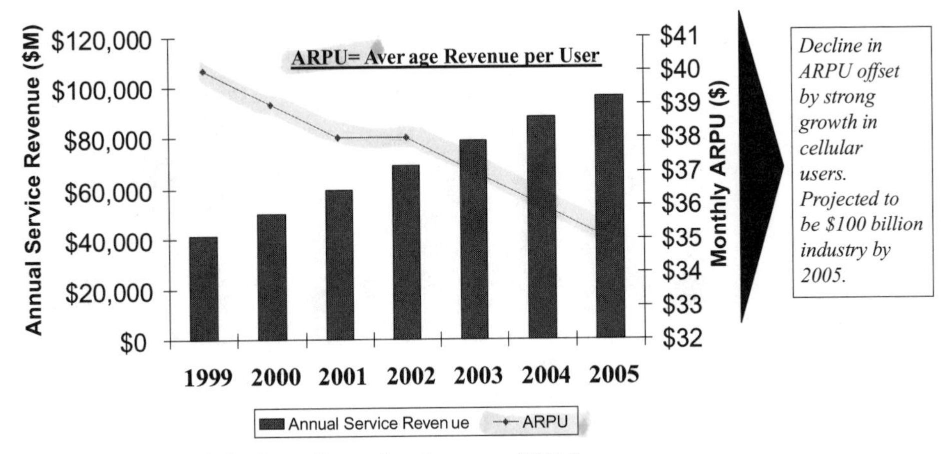

Source: Strategy Analytics, Inc.: a Boston-based company (2000d)

Figure 4.19
U.S. Mobile Wireless Service Revenue and Monthly ARPU.

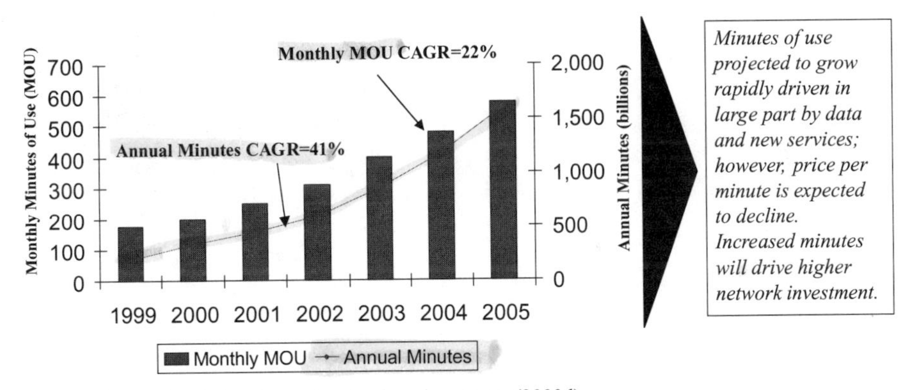

Source: Strategy Analytics, Inc.: a Boston-based company (2000d)

Figure 4.20
U.S. Wireless Minutes of Use.

As a result of increased utilization of the wireless network, both in terms of number of calls placed and MOU, additional capital investment will need to be deployed to increase the capacity of the network, potentially reducing profitability and cash flow.

Carrier Calling Plans

Nextel and AT&T Wireless have traditionally led the mobile wireless communications industry in ARPU. This has been due to their heavy and successful focus on business customers (see Figure 4,22).

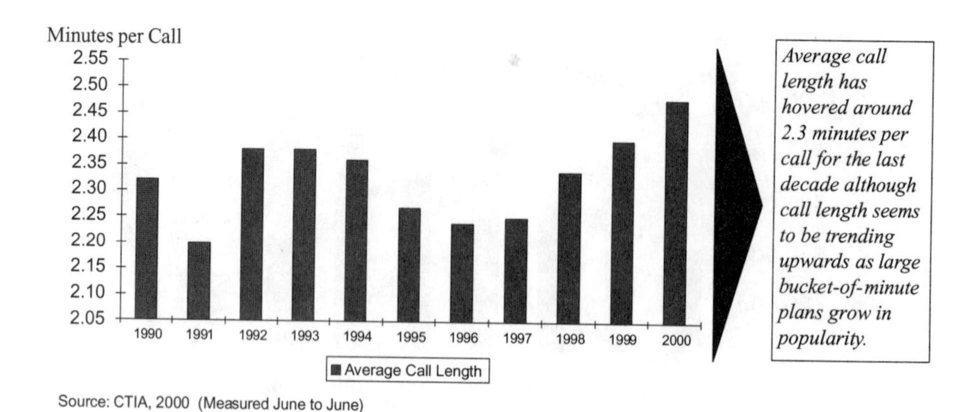

Figure 4.21
U.S. Mobile Wireless Telephone Average Call Length 1990–2000.

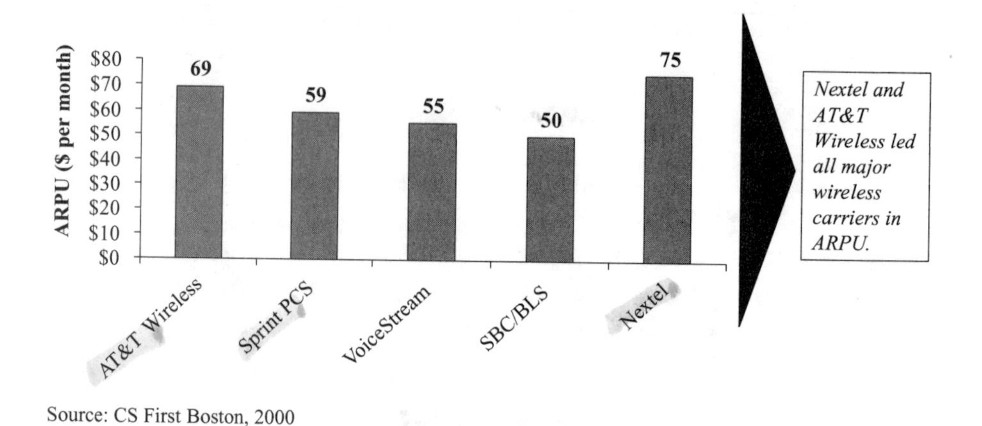

Figure 4.22
Average Monthly Mobile Wireless Revenue per User Third Quarter 2000.

Nextel has used Direct Connect, its proprietary push-to-talk feature, to capture a large portion of the construction industry and other work group occupations. With Direct Connect, also known as Push to Talk, individuals can communicate with their coworkers, at no incremental cost, across a large area.

AT&T, through its ground-breaking Digital One Rate Plan, has been able to win much of the business road warrior market. AT&T's plan guarantees one consistent rate everywhere in the country, with no roaming or long distance charges. The assurance of no roaming charges has been a very valuable feature for customers. However,

AT&T has been paying a tremendous amount of money in network charges to rural carriers to enable this plan. This will probably encourage them to acquire these firms if their level of network usage justifies it financially.

AT&T's Digital One Rate Plan was the first nationwide calling plan without any roaming or long distance charges (Nextel was actually the first to have a nationwide plan). This plan seems to have marked a turning point in the industry. Due to AT&T's strong brand, elimination of any confusion in billing, and heavy promotion, the plan gained a great deal of legitimacy with customers.

Due to the overwhelming success of Digital One Rate, it was copied by many competitors and prices for nationwide service have been falling rapidly (see Figure 4.23). It is important to note that, unlike Digital One Rate, not all plans provide off-network roaming at no additional cost. For example, Sprint PCS charges roaming charges for going off their network. Roaming charges can be substantial and are difficult to estimate.

One of the major benefits of national calling plans is price. It is now often much less expensive to use a wireless phone than some of the traditional substitutes. For instance, pay phones often have very high rates (especially for long distance or collect calls); calling cards often require a call set-up fee of $0.99–$1.99 plus toll charges; and hotel phones are notoriously expensive. Thus, customers are often getting both a substantial discount and far greater convenience with mobile wireless communications.

Provider	Minutes	Price	Per Minute	Provider	Minutes	Price	Per Minute
Verizon	150	$ 35.00	$0.23	Nextel	600	$ 89.95	$0.15
Sprint PCS	180	$ 29.99	$0.17	SBC/BLS	600	$ 89.95	$0.15
SBC/BLS	250	$ 49.95	$0.20	Verizon	600	$ 75.00	$0.13
AT&T	300	$ 59.99	$0.20	Sprint PCS	700	$ 69.99	$0.10
Sprint PCS	300	$ 39.99	$0.13	Verizon	900	$100.00	$0.11
Nextel	400	$ 69.95	$0.18	AT&T	1000	$119.99	$0.12
Verizon	400	$ 55.00	$0.14	Nextel	1000	$129.95	$0.13
Sprint PCS	500	$ 49.99	$0.10	SBC/BLS	1000	$119.95	$0.12
AT&T	600	$ 89.99	$0.15	Sprint PCS	1000	$ 75.00	$0.08

There are a large number of reasonably low-cost national calling plans and prices have been falling.

Source: Company Web sites

Note: As of June 2000 in available service areas

Figure 4.23
National Calling Plans.

Calling Party Pays

One of the anomalies of the U.S. market is the fact that wireless phone customers pay for calls that are both *received* and calls placed. In other parts of the world, such as Europe, the "calling party pays" for wireless calls. A wireless phone number in Europe is identified by a prefix that alerts callers that they are calling a wireless phone and that the call is subject to a charge.

U.S. carriers blame the lack of calling party pays for reducing utilization of wireless communications and hindering the adoption of wireless phones as a replacement for landline telephones. The evidence has shown that customers are hesitant to give their phone number out and often do not turn on their handsets in order to reduce charges. This has suppressed minutes of use and monthly charges. S. G. Cowen estimates that the lack of calling party pays suppresses usage by at least 20% (Cowen 2000, 65). However, this may be becoming less of an issue as low cost, large bucket-of-minute rate plans become the norm.

The Federal Communications Commission is investigating calling party pays and is expected to vote on final rules to give service providers the option of adopting a calling party pays system. The primary concern in switching to calling party pays is how to notify the caller of the charge and the amount. Additionally, roaming charges could substantially increase the cost of a call to a wireless subscriber under a calling party pays system.

Prepaid Plans

A relatively new feature on the wireless landscape is the prepaid plan, which represented at least 28% of the market in 2000 (see Figure 4.24). This is an attractive option for customers who have credit problems and for people who want to regulate spending. As many as 30% of applicants for wireless service in the U.S. are turned down due to impaired credit. This is different from landline telephony where there is a mandate to provide universal service, due to the lifeline status of the telephone.

There are also a large number of people who want to regulate their spending. These include businesses and parents who want to limit teen spending.

Prepaid plans have gained a large share of the market, and this share is consistent with the percent of applicants turned down for credit concerns. The market for prepaid is expected to grow rapidly, both in absolute numbers and in percent of customers (see Figure 4.25). This will be a result of carriers increasingly targeting a younger and less affluent market.

Prepaid is also expected to gain in popularity as prepaid prices become more comparable to traditional plans. There is a substantial premium for prepaid services since acquisition costs are high and revenues are uncertain. This is partially ameliorated by the ability of the carrier to monetize the float on prepaid balances.

AT&T announced a new prepaid pricing plan in time for Christmas 2000 (see Table 4.1). They lowered the prices for prepaid substantially, but there was still a significant premium over the traditional plan.

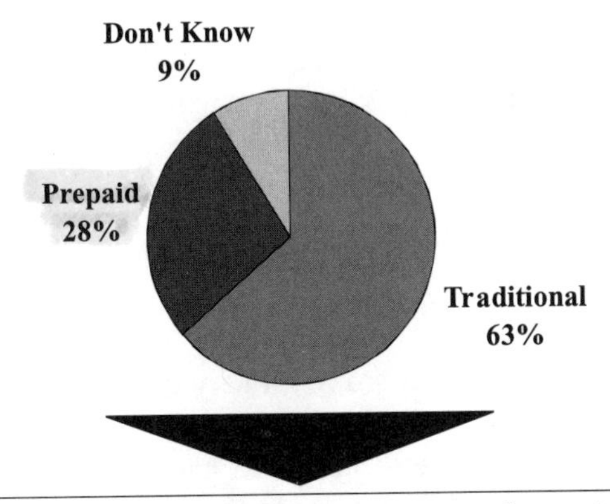

Prepaid plans are a popular means of obtaining wireless service for the credit impaired (approximately 30% of applicants in the U.S. are declined for traditional plans due to credit issues), and those who want to regulate spending (teens, parents, businesses). Prepaid plans are expected to grow rapidly in popularity.

Source: IDC, 1999

Figure 4.24
U.S. Household Subscription Share by Type.

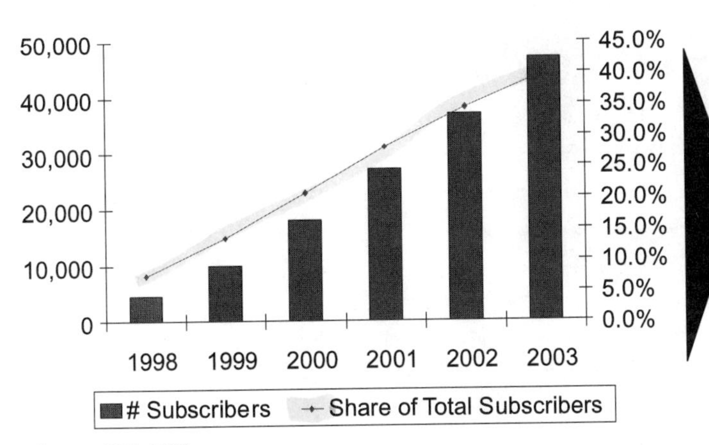

Prepaid subscribers are expected to grow at a CAGR of 58.1%, to 40% of total subscribers as pricing becomes more comparable to traditional plans and carriers target lower income and younger markets.

Source: IDC, 1999

Figure 4.25
U.S. Business and Consumer Prepaid Wireless Subscribers.

Table 4.1 AT&T Prepaid Plans.

Card Denomination	Local Plan Cost/min	National Plan Cost/min
$25	$.35	$.65
$50	$.30	$.50
$100	$.25	$.35
$200	$.15	$.25
Long Distance	Included	Included
Roaming	$.85	Included

Released November 8, 2000. Source: AT&T Web site

Service Channel

Using the source of handset purchase as a proxy for service acquisition channel, one can see that the carrier is clearly the dominant channel for wireless service (see Figures 4.26 and 4.27). Since service activation and handset purchase are usually intertwined, due to handset subsidies, this is a very good proxy for service decisions. Historically, fourth quarter holiday sales have been a strong driver of sales and this will probably continue as the prepaid and youth markets are more actively targeted (see Figure 4.28).

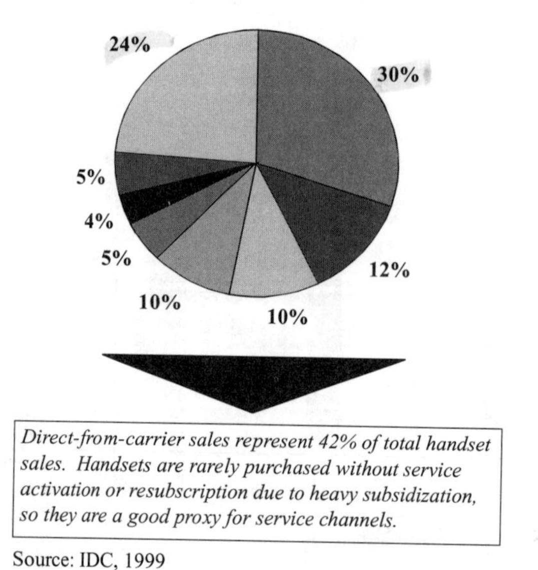

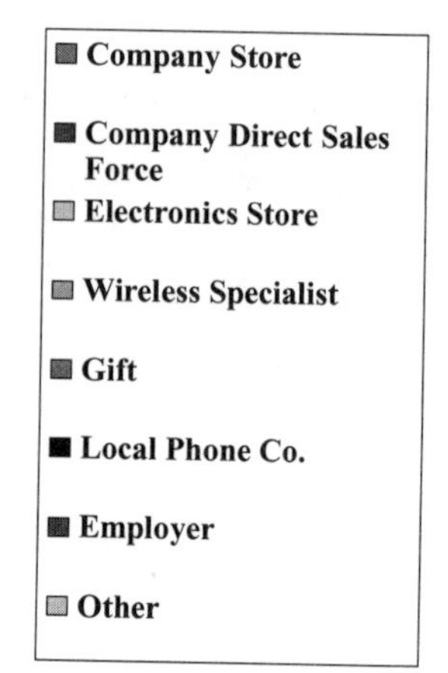

Source: IDC, 1999

Figure 4.26
Source of Most Recent Handset Purchase.

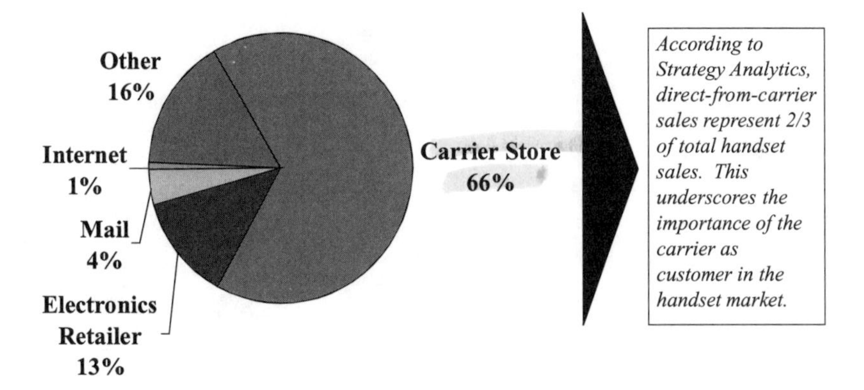

Source: Strategy Analytics, Inc.: a Boston-based company (2000c)

Figure 4.27
U.S. Handset Distribution.

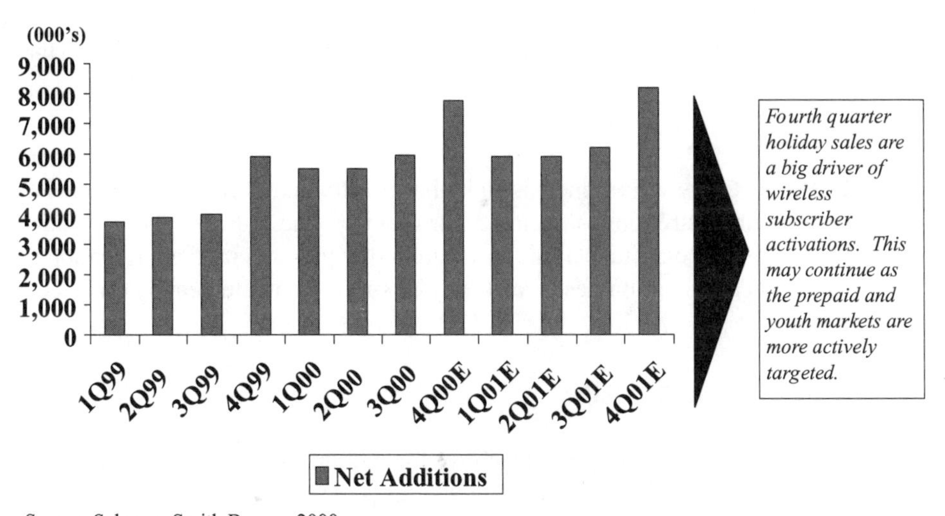

Source: Salomon Smith Barney, 2000

Figure 4.28
Net Mobile Wireless Subscriber Additions by Quarter.

Structural Market Assessment

For the foreseeable future, the market structure of the mobile wireless communications carrier market will be extremely challenging (see Figure 4.29). This is likely to lead to intense competition and foster rapid consolidation (government permitting). However, the competitive landscape may change once the industry is able to consolidate to two or three major carriers. As in many industries, there may not be room for more than a few major players in mobile wireless communications.

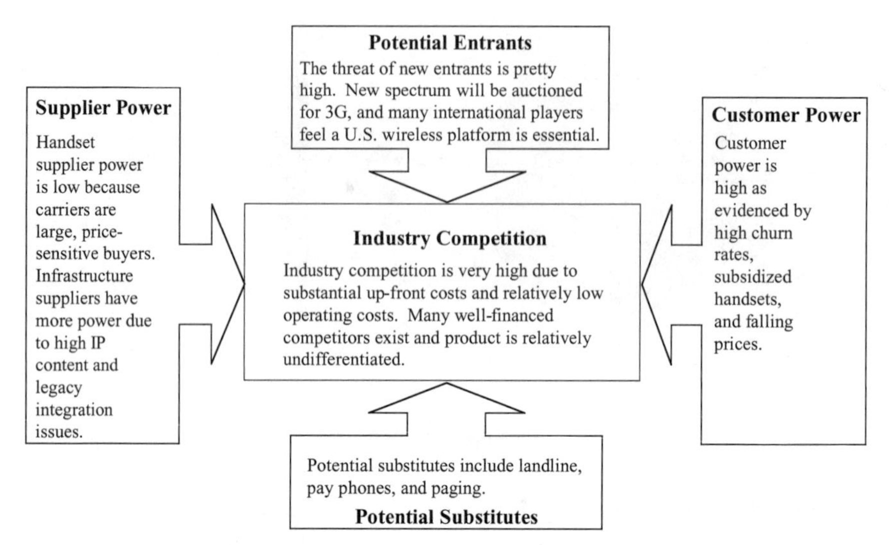

Figure 4.29
Structural Market Assessment: Voice (Data) Network Service Providers.

Customer Power

Customer power is extremely high in the industry and is evidenced by the lack of profitability and astronomically high churn rates in the industry. The customers clearly view wireless communications as a commodity product, particularly among the megacarriers. This situation has forced the industry to compete heavily on price.

Potential Substitutes

There are several potential substitutes for mobile wireless telephony. The main ones are traditional landline phones, pay phones, and paging. The existence of these generally lower cost options is one of the reasons that wireless did not become a mass market item until prices dropped dramatically. However, with much more competitive pricing, the industry has gained mass market acceptance. In fact, for travelers, nationwide plans are now much less expensive than using pay phones or phones in hotel rooms. With large bucket-of-minute plans offered at reasonably low fixed prices, landline replacement has emerged as a real possibility.

Supplier Power

Supplier power is relatively low in the industry, with the exception of the supplier of spectrum—the U.S. government. The government has a great deal of power in the industry, including the power to enable new competitors. Handset suppliers are weak compared to carriers, and this is evidenced by their low margins and increasingly competitive market. Infrastructure suppliers are more powerful than handset suppliers, but are less powerful than carriers. The key strengths of the infrastructure suppliers are their high intellectual property content and their ability to integrate high-quality equipment into a legacy network environment. Carriers are able to apply their power against infrastructure suppliers through large contracts awarded through a very competitive bidding process between several strong and capable suppliers.

Potential Entrants

In 2001, the threat of new entrants is reasonably high as new spectrum is scheduled to be auctioned by the FCC and large foreign carriers seek to enter the U.S. market as part of their strategy to develop a global platform. Vodafone has established a substantial presence in the U.S. through their holdings in Verizon Wireless, Deutsche Telekom is attempting to enter the U.S. market through its proposed acquisition of VoiceStream, and NTT DoCoMo acquired 16% of AT&T Wireless in December 2000.

Even in the face of limited returns, more competitors are attempting to enter the market because they fear being left out more than they fear poor returns. The threat of new entrants may go down as businesses more accurately assess the potential returns of the industry under the existing market structure.

Industry Competition

The competitive intensity of the industry is extremely high due to the large number of competitors and the nature of the investment. Investment in a wireless network requires a huge up-front investment in spectrum licenses and infrastructure equipment, while network operating costs, with the exception of customer acquisition costs, are relatively low. This is consistent with the economics of most telecommunications networks. The economic conditions faced by network operators compel them to fight viciously for new customers. With at least four competitors in every market, this level of competition can be quite destructive, especially to profits. Further fueling this competition is the positive price elasticity that mobile wireless communications has exhibited over the past decade. This encourages carriers to lower prices in order to gain or maintain share.

This economic situation provided the rationale for telephone monopolies in the past. However, Congress and the FCC have decided that competition in communications is in the best interest of the country, from both a pricing and innovation standpoint. There is clear evidence that as competition in telecommunications increases, prices decline, often precipitously. The long distance industry provides a good example of this dynamic.

Customers recognize this situation and have capitalized on the weakness of the carriers, demanding—and getting—steep price cuts. The carriers do not presently have any unique levers at their disposal to increase customer loyalty. Thus, carriers are continually searching for strategies to create personalized services in order to retain customers.

AT&T and Nextel have done the best job of differentiating their products with their Digital One Rate Plan and Direct Connect services, respectively. Digital One Rate allowed AT&T to capture a large share of the business traveler market and Direct Connect gave Nextel a strong offering for work groups. This is reflected in both companies' relatively high ARPUs. However, AT&T's competitors have been able to copy their nationwide plan and prices have been falling. Nextel's proprietary system remains unique, but the overall pricing environment impacts them as well.

Thus, we have a very competitive industry that is growing rapidly in terms of both subscribers and revenues. There are also substantial new revenues on the horizon as data becomes a larger proportion of network traffic. The main problem, however, is that this may be a profitless prosperity. Given this situation, one would expect the industry to undergo further consolidation. This has been happening, and consolidation may accelerate in the future given the economic circumstances. The one fly in the ointment may be the government, who may not favor consolidation after all their efforts to create competition. The government enforces the level of competition through spectrum caps that ensure there are at least four competitors in every market.

Outlook

The wireless carrier market is an enormous market growing extremely rapidly, especially for its size. This market will continue to grow rapidly, for the foreseeable future, driven by increased subscribership and minutes of use. This will persist as prices fall and the value proposition of mobile wireless communications is increasingly recognized by subscribers. Subscribers will also continue to substitute wireless telephony for landline telephony, as is the case with long distance calls, where calling patterns have already been impacted.

The advent of broadband, packet-switched networks and value-added data services will provide another leg up for carriers and unleash some truly marvelous capabilities. However, substantial investments by the carriers will be required to enable these services.

Unfortunately for the carriers, the industry dynamics are very unfavorable and likely to result in hypercompetition. Consolidation would be the natural path for this industry to follow. However, the FCC may retard the industry's ability to follow this path through its spectrum caps.

The long-distance market provides a good analog to wireless telephony, in that there are a number of undifferentiated suppliers of a commodity product that requires a large network investment with low operating costs and high customer acquisition costs. The returns in the long-distance business have been eroding for years, and the same can be expected in wireless communications if consolidation is blocked.

MOBILE WIRELESS DATA/INTERNET . . . .

This section addresses wireless data transfer and Internet access over cellular networks rather than over paging/data networks. Paging/data networks are discussed separately in Chapter 2.

Market Size and Forecast

Mobile wireless data/Internet is the transfer of nonvoice data in a mobile wireless environment. This data can include text, images, audio, video, or large files. Mobile wireless data and Internet access represent the next major growth driver for the wireless industry, and these services are counted on heavily by wireless carriers. This is evident in the bidding seen for 3G licenses in Europe (see Chapter 1).

Mobile wireless data/Internet represents the convergence of two of the fastest growing technologies of the late twentieth/early twenty-first centuries—or any other time for that matter—wireless communications and the Internet. The adoption rates for the Internet and wireless communications are breathtaking when compared with the adoption rates of other, now ubiquitous, technologies. According to Commonwealth Associates, it took the mobile wireless phone 15 years to reach 25% market penetration, and only 7 years for the Internet to reach this level (Greiper and Ellingswoth 2000, 2). In contrast, it took 35 years for the landline telephone to cross this threshold.

The convergence of these two technologies, in conjunction with location technologies and Bluetooth, represents one of the most exciting and important business opportunities in history. These technologies can truly enable instant access to virtually any piece of information from anywhere. This information can be provided in a proactive fashion within a location-specific context, and can then be transmitted to a correspondent device through a Bluetooth transceiver. This is a cataclysmic shift in personal capability and is virtually unparalleled in history, with the possible exception of the freedom enabled by the automobile. This increase in personal capability should not be underestimated and will change the way people live and work in ways we can't imagine today.

Mobile wireless Internet is projected to grow explosively in terms of subscribers, penetration, and the amount of data transferred; see Figures 4.30 and 4.31. Note the difference between the two figures' forecast periods. High forecast for 2003 in Figure 4.31 was 35+ million from IDC. However, it is important to be aware that there are many forecasts regarding the growth of mobile wireless data/Internet. These forecasts vary greatly, and should be interpreted with a high degree of caution. There

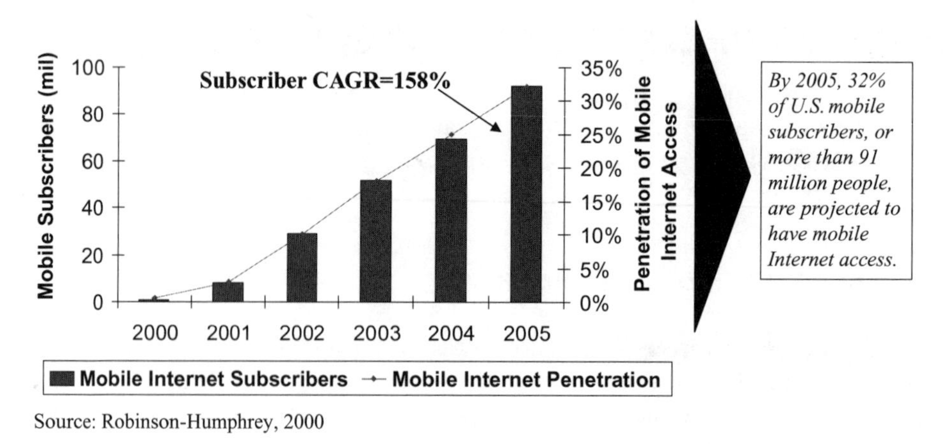

Source: Robinson-Humphrey, 2000

Figure 4.30
U.S. Mobile Wireless Internet Penetration.

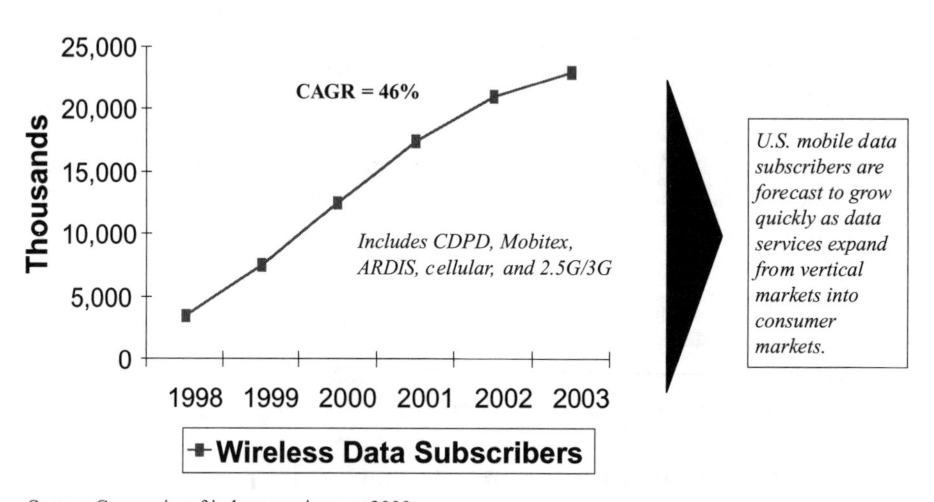

Source: Composite of industry estimates, 2000

Figure 4.31
U.S. Mobile Data Subscribers.

is certainly an enormous market opportunity in mobile wireless data/Internet, but there are also many unrealistic expectations in the industry and a number of key hurdles remain before widespread adoption is assured. These hurdles include the availability of powerful, reasonably priced terminals; the deployment of high-speed, packet-switched networks; and compelling content. The excitement surrounding mobile wireless data/Internet is very reminiscent of the big push in the mid to late 1990s to market goods to China. Everybody was rabidly excited by the prospect of a billion new customers to outfit, but less attention was paid to the enormous challenges in successfully accessing them.

There are also a number of forecasts regarding different wireless data and multimedia services that will be adopted and their resulting revenues. It is useful to consider the primary purpose of the wireless handset when assessing the probability of these estimates—mobile communication. Voice communications is the killer app of the wireless phone (the overwhelming majority of data access devices will be wireless handsets), and communication is likely to be the primary data service used by subscribers. This will be in the form of SMS or email messages.

Other services likely to be adopted initially involve general information—news, weather, and stock quotes. Location-based services, when they are widely available, should also prove to be very popular with consumers. Video and audio streaming, while capturing the imagination of many, are apt to develop much later as costs come down, content is developed, and a genuine need is perceived by customers.

It is important to remember that messaging and basic information are virtually free services in the wired world, and this is also likely to be the case in the wireless world. These services should be expected to be included in most flat-rate pricing schemes from carriers, and in the long run may not add much to the carrier's bottom line. Therefore, in order to prosper, service and content providers will need to develop more value-added services targeted toward the enterprise market.

As a further note of caution, many market research studies have found that there is not an overwhelming unmet consumer demand for mobile wireless Internet services in the U.S. This is likely to necessitate a major push by network service providers to educate customers about the benefits and applications of the service. This will also have implications with regard to pricing.

In 2000, when mobile wireless Internet access was in its infancy, Morgan Stanley estimated that there were 735,000 U.S. mobile wireless Internet users on browser-embedded phones as of September 2000 (Lundberg and Zucker 2000b, 45).

Wireless versus Wireline Internet Access

There are substantial and compelling benefits in accessing the Internet from a mobile wireless device versus a desktop PC. These include lower cost (device), portability, availability (device can be with a person 24 hours a day), and location information. The drawbacks to mobile wireless data devices include small screen size, limited input methods, limited processing power, limited battery life, uneven network coverage, and unproven reliability and security. The advantages of mobility are undeniable, and the drawbacks are technical in nature and will be successfully addressed through new and improved technologies. However, this process will take time, potentially until the 2003–2005 timeframe, before most of these issues are successfully addressed.

As mobile wireless Internet subscribership grows rapidly, data is likely to surpass voice traffic on cellular networks sometime around 2004–2005, as it has already done on the wireline telephony network (see Figure 4.32). It is also likely that, in this time frame, the number of individuals employing wireless Internet access will surpass those using wireline access (see Figure 4.33). Robinson-Humphrey (p. 15) projects that more than 50% of Internet hits will originate from wireless devices by 2004. This will be enabled by broadband, packet-switched 2.5G and 3G networks, which are scheduled to begin deployment in 2001.

Packet switching will be tremendously valuable for wireless data services, as spectrum is used much more effectively and bandwidth increases. Additionally, the "always on" nature of packet-switched networks does not require the customer to dial up in order to establish a connection. This will allow timely and pertinent information to be pushed proactively to the subscriber.

One of the keys to wireless data will be successfully navigating the myriad of interface requirements, differing transmission technologies, and wide variety of terminals and displays.

Subscribers will want to use many different types of terminals to access data from a variety of different data sources, including the company mainframe and the

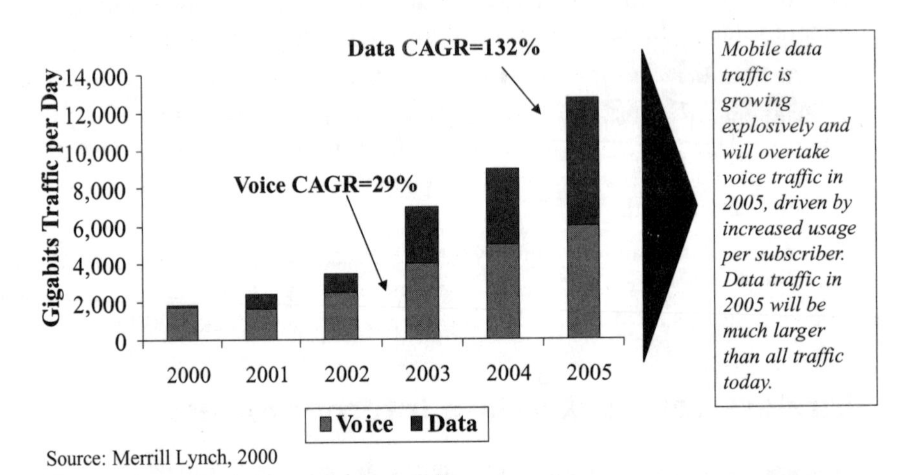

Figure 4.32
Mobile Voice versus Data Traffic Worldwide.

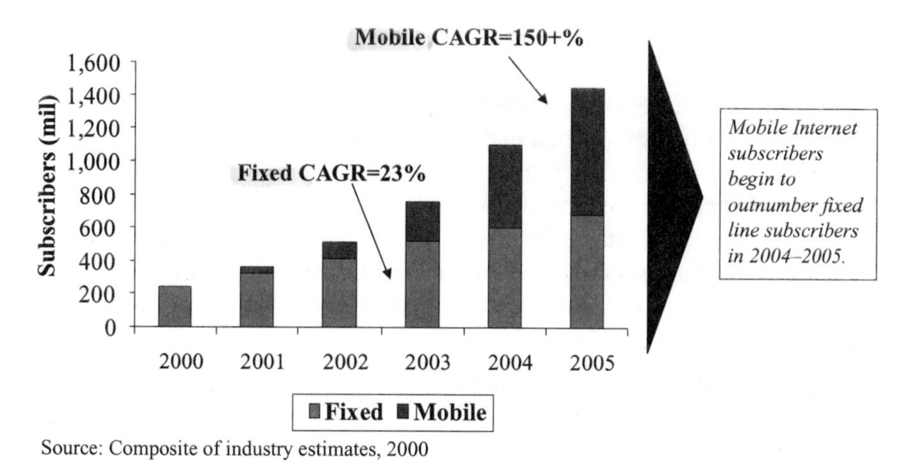

Source: Composite of industry estimates, 2000

Figure 4.33
Mobile versus Fixed Internet Users Worldwide

Internet. They will also want to be able to roam off of their primary service provider's network. These types of differing interfaces will all need to be successfully managed and supported to fully enable mobile wireless data and Internet.

The roaming issue is a substantial challenge that does not receive adequate coverage in the popular press. The ability to provision and properly bill for packet-switched services across various networks in the U.S. is a daunting challenge. Both the infrastructure and billing agreements will need to be in place throughout the network before roaming services can be offered.

In the U.S., the three main transmission technologies are CDMA, TDMA, and GSM (see Chapter 1). These systems employ substantially different data-transmission schemes and they will evolve along several different paths toward a 3G solution.

There are also a wide array of wireless terminals, including handsets, PDAs, pagers, and laptop computers (see Chapter 3), which people will want to use to access data. These devices have differing screen sizes and formats, different modems, and different operating systems. The requested data will need to be served in an appropriate manner for each of these devices.

The confluence of these circumstances will require a system capable of accommodating a data request to the company inventory server from Palm V across a CDMA network, as well as a request from a Motorola handset, with four-line text capability, operating on a CDPD network, and having the reply be intelligible to both requestors. This is no small feat, and will require a number of enabling gateway technologies. A gateway is generally a server and software (often called middleware) that translates data between systems. These gateways and middleware will be essential for mobile wireless data.

Provisioning Wireless Data

Data suppliers have two main options in provisioning data to wireless subscribers. They can supply preformatted data to the carrier's network for the wireless device, or they can supply their data to the network service provider and rely on the carrier to properly format the data. Additionally, a wireless hosting company can be used; this is similar to preformatting the data. Providing preformatted data to the network has been the preferred solution, since it gives the content provider more control over the look and feel of the data and allows them to provision data across networks that do not have a mobile server.

There are a number of cellular data services available in 2000. These include SMS, WAP, and CDPD (these are in addition to the paging, Mobitex, and ARDIS networks that are discussed in the Chapter 2). See Figure 4.34.

SMS

Short Message Service (SMS) is the ability to send and receive text messages from a wireless telephone. It was first used in 1992 and is part of GSM standard. SMS can send/receive up to 160 characters (70 characters non-Latin alphabet) from an SMS-enabled wireless handset.

The most common use of SMS in the U.S. in 2000 is for voicemail notification on wireless phones. When a screen says there are "3 voicemail messages," it is an

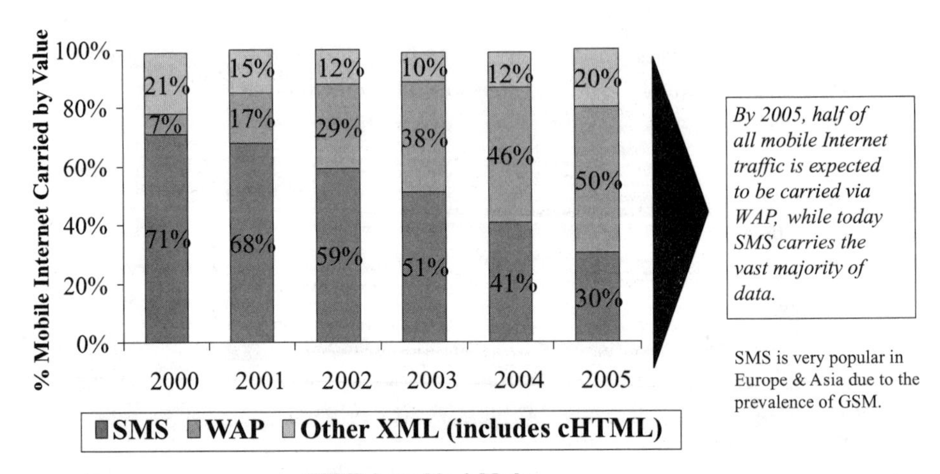

cHTML is used by i-Mode

Source: The European Telecom Team, Merrill Lynch. Reprinted by permission.

Figure 4.34
Wireless Mobile Internet by Transport Protocol Worldwide.

example of a one-way SMS message from the wireless carrier. This feature does not generally work when a subscriber is roaming off network.

SMS has become wildly popular in Europe, generating more than 1 billion messages per month and accounting for up to 10% of revenues. The typical charge for sending an SMS message in Europe is $0.15. SMS is particularly popular with the youth market and is used to notify each other of rave party locations and to pass notes during class.

Other potential SMS applications include unified messaging and notification, chat, information services (push and pull), m-commerce, customer service, vehicle positioning, job dispatch, and credit card authorization.

WAP

Wireless Application Protocol (WAP) is a global standard designed to make Internet services available to mobile users by converting Internet content to a format suitable for the limitations of mobile handsets, such as reduced screen size, lack of a keyboard, and limited memory and processing power.

The WAP standard was developed and heavily supported by the WAP Forum (see Figure 4.35). The primary founders of the WAP Forum were Openwave Systems

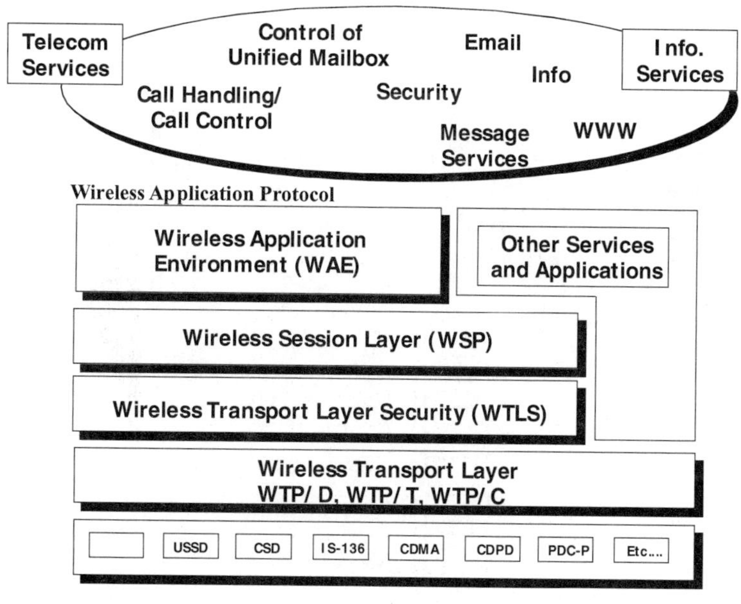

Source: IDC, 2000

Figure 4.35
WAP Infrastructure Overview.

(formerly Phone.com), Ericsson, Motorola, and Nokia. The WAP standard has been adopted by more than 90% of the wireless carriers worldwide.

The WAP protocol uses wireless mark-up language (WML), and is more efficient at delivering information than HTML and other similar languages. WAP is text based and menu driven, and in 2000 did not support color or graphics. To fully implement WAP on a network requires a WAP server, a separate WAP Web site written in WML, and a WAP-enabled handset with a WAP browser (see Figure 4.36).

WAP Version 1.1 has been developed to work with circuit-switched networks; WAP Version 1.2 will operate on packet-switched networks.

Openwave is the dominant supplier of WAP servers and browsers, and has 85 of the largest wireless carriers, worldwide, as customers (see Figure 4.37). Nokia also markets a WAP server and Ericsson is planning to enter the market in 2001.

Initial results of WAP services in Europe have been characterized as disappointing due to slow speeds, high costs, and limited content. However, it is important to remember that these are network and content problems, not problems with WAP itself. WAP is simply the protocol, not the content, and it is reliant on the network for transport speeds. The term WAP, though, has seemingly come to encompass the entire industry. This is typical of the confusion between terms and meanings in the emerging wireless market.

Looking forward, WAP services should improve dramatically as higher-speed packet-switched networks are implemented.

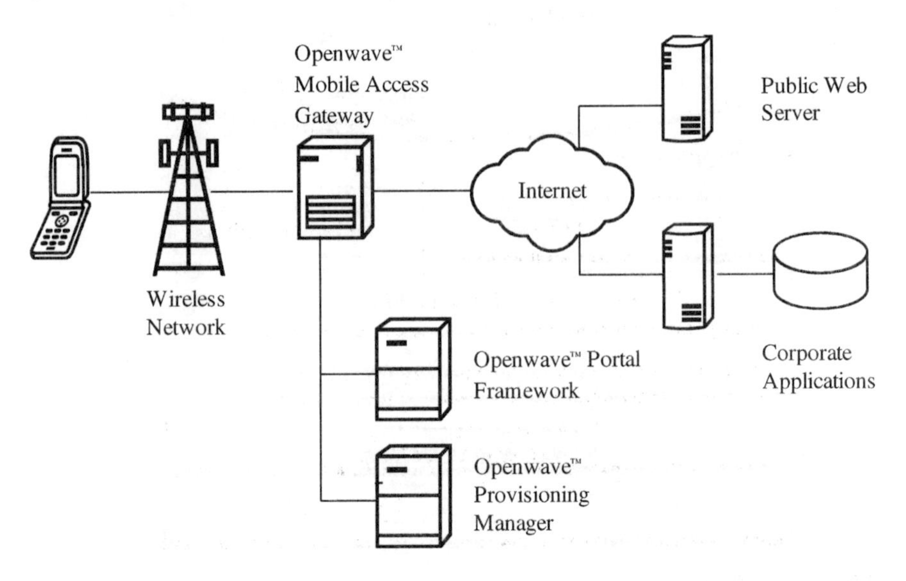

Source: Openwave

Figure 4.36
Openwave's Wireless Internet Solution.

CDPD

Cellular Digital Packet Data (CDPD) is a packet-switched data transmission technology developed for use on cellular frequencies, and is considered a precursor to GPRS. CDPD uses unused cellular channels (in the 800 MHz to 900 MHz range) to transmit data in packets, and offers data transfer rates up to 19.2 kbps.

AT&T's PocketNet service and the Novatel and Sierra PDA modems use CDPD. Other carriers that support CDPD include Verizon, ALLTEL, and Ameritech. CIBC estimates that there were 600,000 individuals using CDPD services in the third quarter of 2000. Roughly half of these were AT&T PocketNet subscribers, who numbered about 300,000 in the third quarter of 2000.

i-Mode

i-Mode is a mobile wireless data service from NTT DoCoMo that was launched in Japan on February 22, 1999, and signed up more than 12 million subscribers in the first 18 months (see Figure 4.37). As of August 2000, 591 companies were providing information services through i-Mode, which was a huge increase from the 67 firms that were signed on when the service launched. Additionally, there were about 1,000 official i-Mode Web sites and another 18,700 independent sites, according to OH! NEW i-search, an i-Mode search engine.

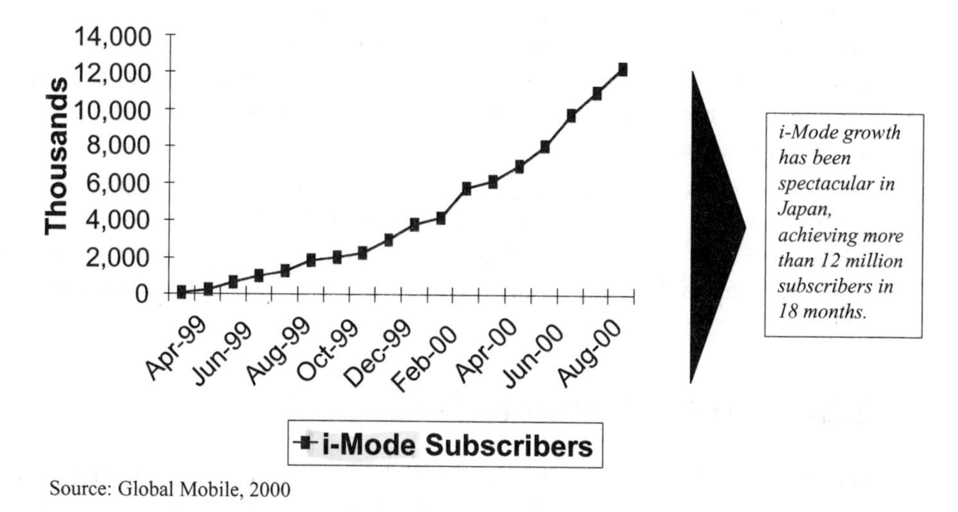

Source: Global Mobile, 2000

Figure 4.37
NTT DoCoMo i-Mode Subscribers.

i-Mode allows subscribers to reserve airline and concert tickets, check their bank balances or transfer money, send and receive email, and access the Internet directly from their i-mode-compatible cellular phone.

i-Mode uses the Compact HTML (cHTML) language and a packet-data (9.6 kbps) transmission system, where customers are charged according to the volume of data transmitted, not the time spent online. Although 9.6 kbps is slow, the packet-based nature of the network, strong content offerings, and lack of a strong wireline Internet presence in Japan have allowed i-Mode to be spectacularly successful. Japan will also be the first to roll out 3G services when NTT DoCoMo launches its W-CDMA network in 2001.

i-Mode has been a huge financial boon for DoCoMo. DoCoMo charges a $3 per month subscription fee for access to the service and charges for optional services, which averaged about $5 per month. There is also a charge for network usage, estimated by Morgan Stanley to average $12 per month in 1999 (Lundberg and Zucker 2000b, 47). The total lift in ARPU per i-Mode subscriber is approximately $20 per month, or $5 per month across all subscribers. NTT DoCoMo also collects a 9% fee from any e-merchants that use their integrated billing structure.

i-Mode has been a tremendous success in the Japanese market, and many industry observers think that this will translate to the U.S. market. There are, however, a number of key differences between the U.S. and Japanese markets. PC Internet penetration is extremely low in Japan, at around 10%. This is due, in large part, to the high expense of accessing the Internet from a landline phone in Japan. In 2000, Internet access charges could easily run $100 per month, when the ISP charge and metered phone service charges were totaled. Therefore, for many Japanese, i-Mode is their first Internet experience. This impacts their perception of the Internet, since they are not accustomed to high-speed Internet connections and 17-inch color monitors, as is the case in the U.S.

Although i-Mode was strictly a Japanese phenomenon in 2000, it may well migrate to the U.S. NTT DoCoMo has acquired 16% of AT&T Wireless, which may provide a U.S. beachhead for i-Mode. This relationship will almost certainly lead to coordinated technology road maps and international roaming between the U.S. and Japan.

2.5G and 3G Technologies

The 2.5G and 3G packet-switched technologies will enable many new mobile wireless data services and applications. These applications will include high-speed data transfer and video and audio streaming. This will allow the mobile Internet to really come of age.

GSM/TDMA Path

There are two main paths to 3G. One is from the TDMA/GSM camp that can follow the GPRS to EDGE to W-CDMA (steps can be skipped) path and the other is from the CDMA camp using the CDMA2000 migration path with 1X, and its derivatives, to 3X. These standards aspire to the advertised 3G speeds of 2 Mbps in a stationary environment and 384+ kbps in a mobile environment.

GPRS—General Packet Radio Service (GPRS) is a 2.5G technology that allows networks to send "packets of data" at rates up to 115 kbps. GPRS allows "always on" connections to send information immediately to the subscriber, with no dialup required. GPRS is more efficient than sending data over a circuit-switched wireless connection and will allow users to be charged per packet of data rather than by connection time.

GPRS is a data-only packet network overlay for GSM networks and is a relatively straightforward upgrade to existing networks, requiring a software and chip board upgrade. GPRS also requires a Gateway GPRS Support Node (GGSN), which is a packet router, and a Serving GSN Support Node (SGSN), which tracks the subscriber and provides security.

In 2000, the only handset available for GPRS services was from Motorola, and only in limited quantities. While 115 kbps is the advertised speed for GPRS, handsets have not yet been able to achieve these speeds. Apparently, heat generation has been a serious problem, even at much slower speeds. GPRS handsets are expected to be more widely available in the second half of 2001, when Nokia is scheduled to release its product.

AT&T Wireless has announced it will begin to roll out a new GSM/GPRS network in 2001.

EDGE—Enhanced Data Rates for GSM Evolution (EDGE) is an evolutionary path to 3G services for GSM and TDMA operators. It represents a merger of GSM and TDMA standards and builds on the GPRS air interface and networks.

EDGE is a data-only upgrade and supports packet data at speeds up to 384 kbps. It is able to achieve increased data transmission speeds through a change in its modulation scheme, from GMSK to 8 PSK. The upgrade to EDGE is relatively expensive and requires carriers to replace the transceivers (radio antennas) in every cell site. According to Yankee Group this can cost as much as 60% of the original network cost (Greiper and Ellingswoth 2000, 21).

AT&T Wireless has indicated that they plan to migrate to EDGE.

W-CDMA—Wideband Code Division Multiple Access (W-CDMA) is supported by GSM operators as their 3G technology, and is viewed as the best evolutionary path for GSM/TDMA operators. W-CDMA will support speeds of 384 kbps to 2 Mbps and will use a faster chip than CDMA2000. W-CDMA will not be as compatible with early versions of CDMA as CDMA2000, since the W-CDMA chip is set up for the timing of GSM rather than the timing of CDMA.

W-CDMA uses a wide channel, 5 MHz, and requires a complete network overhaul. The installation cost for W-CDMA is estimated by Commonwealth Associates to be 100–120% of the original network cost.

NTT DoCoMo, Japan's largest wireless carrier, is expected to be the first global carrier to implement W-CDMA—in 2001—when it rolls out 3G services in select Japanese metropolitan markets. AT&T Wireless has committed to W-CDMA, with probable deployment in 2003–2004, to enable international roaming with their partner NTT DoCoMo. NTT DoCoMo purchased 16% of AT&T Wireless in November 2000.

CDMA Path

CDMA2000—CDMA2000 is an evolution to 3G technology for CDMA networks. CDMA2000 is compatible with current CDMA networks, IS-95A and IS-95B (data-enabled CDMA).

CDMA2000 is expected to be rolled out in two phases, 1X or 1XRTT and 3X or 3XRTT (RTT stands for Radio Transmission Technology). 1XRTT is expected to provide a packet data rate of 144 kbps in a mobile environment and operate on 1.25 MHz of bandwidth (1X). It is also forecast to lead to a twofold increase in voice capacity and battery standby time. Commercial availability is expected in 2001.

1XRTT is a relatively straightforward upgrade requiring new software and a chip board. According to Yankee Group this is likely to cost about 30% of the original network cost, which is substantially less than the cost of upgrading other networks, and has the advantage of substantially increasing voice capacity (Greiper and Ellingswoth 2000, 17). Sprint PCS and Verizon have announced they will use 1XRTT. Sprint plans to roll out 1XRTT in late 2001 and expects the cost to be $700–$1,000 million.

Additional intermediate CDMA solutions have also emerged. These include 1XEV, also known as High Data Rate (HDR), and 1Xtreme. 1XEV, from Qualcomm, is estimated to support transmission speeds in excess of 2 Mbps. 1Xtreme is a proposal from Motorola and Nokia that is reputed to support peak speeds over 5 Mbps and steady state speeds of 1–2 Mbps.

3XRTT promises data rate increases to 2 Mbps and will use 3.75 MHz of bandwidth (3X). Commercial availability for 3XRTT is expected in 2002. However, with the emergence of higher-speed intermediate solutions, the need to move to 3XRTT may be reduced. No carrier has committed to 3XRTT as of January 1, 2001.

Data Carriers and Services

Wireless data/Internet represents a tremendous opportunity for carriers and enabling firms. Data is the single best opportunity for carriers to increase network usage and ARPU. Data will also allow carriers to create personalized services for subscribers that should lead to increased customer loyalty and reduced churn.

Carriers have a substantial advantage in the wireless Internet and will be loath to cede their pole position—as they did with the wireline Internet—to companies like AOL and Yahoo. The key service provider advantages include the ability to provide phones with their own portal as the default and the ability to restrict changing the default portal. Carriers will also determine which technological upgrades will be made, and where and when they will occur. This will allow them to ensure that their mobile Internet services are optimized for present and future wireless network systems. The carriers are also able to influence the speed at which WAP or other data services are taken up via handset subsidies and incentives.

The carriers begin the battle for the mobile wireless Internet with another huge advantage, the customer relationship. They already have a very large installed base of customers. Thus, they will attempt to expand an existing relationship with their customer, rather than having to spend several hundred dollars to acquire a subscriber. Controlling a large number of subscribers also allows service providers to negotiate a higher share of revenues from Internet transactions and to obtain substantial slotting fees from content providers. Slotting fees are the fees paid by content suppliers to secure premium placement and positioning on a portal or a screen-based menu. Appearing as the first choice on a menu is extremely valuable to content providers, as usage drops rapidly as more keystrokes are required.

Carriers are also responsible for customer care and thus maintain the critical billing relationship. This can allow carriers to provide integrated billing for merchants as DoCoMo does with i-Mode. DoCoMo collected a 9% commission for providing this service in 2000.

Carriers will also, most likely, have knowledge of customer location for location-based services, and unless forced by law, may be unwilling to share this information without a charge. Having location information will enable a multitude of custom, high value added services to be provided by the carriers.

In 2000, every major carrier in the U.S. market had both a data offering and a text messaging offering available to their customers (see Tables 4.2, 4.3, and 4.4).

Even though carriers have a strong starting position in wireless data services, a unique feature of the U.S. market is the extremely well-developed wireline Internet. This is a key difference between the U.S. and both Europe and Asia. In the U.S., wire-

Table 4.2 Carrier and Online Venture Mobile Wireless Data Partnerships.

Carrier	Portal	News Special Content	Financial Services	Commerce
Sprint PCS	AOL Earthlink Yahoo! Infospace	Go2.com CNN.com Weather.com Bloomberg.com Foxsports.com Mapquest Dictionary.com Emazing.com Afronet	Ameritrade.com Harris Bank	Amazon.com eCompare.com GetThere.com eBay
AT&T PocketNet	Infospace	ABC News Bloomberg.com ESPN.com Zagat.com	E*Trade DLJDirect	BN.com (Barnes&Noble) FTD.com
Verizon	Infospace	Bloomberg.com	MSN Money Central	Amazon.com
Nextel Online	MSN	MSN Strategy.com	N/A	Amazon.com

As of April 2000. Source: Jupiter

Table 4.3 Mobile Wireless Internet Services.

Provider	Service	Price	Features
Verizon Wireless	Web Access (CDMA)	Web Access is $6.95 per month in addition to regular calling plan.	Web Access allows users to access Internet-based information services from more than 20 providers including news, sports, stock quotes, weather, and travel information.
	Web Access Plus (CDPD)	Web Access Plus service begins at $24.95 per month for Palm users; prices higher for laptop users.	Web Access Plus offers wireless Internet connectivity for computing devices such as laptops, PDAs, etc.

Table 4.3 Mobile Wireless Internet Services (continued).

Provider	Service	Price	Features
AT&T Wireless	PocketNet (CDPD)	Basic Plan free with AT&T AT&T Digital Calling Plan. Plus Plan is $6.99 per month. Premium Plan is $14.99 per month.	PocketNet offers users a broad array of content choices including news, sports, travel information, driving directions, Yellow/White pages with auto-dialing, and email access.
Sprint PCS	Wireless Web (CDMA)	Sprint PCS offers three pricing options:	Wireless Web allows users to visit specially designed sites for
		Wireless Web is a free option at sign-up, allowing voice minutes to be used for Wireless Web calls.	news, weather, financial data, shopping, etc.
		Wireless Web service can be added for $9.95 per month to an existing service plan of $29.99 or more, allows you to use calling plan minutes for both voice and Wireless Web calls, includes 30 Wireless Web updates. Each additional minute costs $0.25 per minute.	Wireless Web updates deliver news, sports, weather, and stock quotes directly to the subscriber's phone at times selected by the user.
		Users may access the Wireless Web service at a rate of $0.39 per minute.	
Nextel		$9.99 per month for basic package.	Both packages offer unlimited use of MSN Mobile, Shopping, and Nextel services, with no additional cost or deduction from voice minutes.
		$39.95 per month for basic package with Web browsing.	The Web browsing option allows the user to browse the Internet from the Nextel phone, access hundreds of Web sites, and bookmark the most frequently visited sites.

Table 4.4 Mobile Wireless Text Messaging Plans.

Provider	Price	Services	Limits/Features
AT&T Wireless	Offered as part of calling rate package.	Unlimited use of email dispatched messaging and AT&T's online Messaging	Messages are limited to 150 characters or 230 characters for customers with PocketNet service.
	MessageFlash Software package available at $9.95.	Web-dispatched text messaging. The Web site allows users to maintain address books.	The system will store and resend unsuccessfully delivered messages for up to 3 days.
	Additional fees apply for operator assisted messaging.	MessageFlash software allows users to maintain individual and group directories, store sent messages, dispatch messages at specified times.	
		With Outcall Notification, users through a link with PBX can be notified when a message is left in the users' office voice mailbox.	
		With AT&T Digital PCS Complimentary Personal News, Web updates of news, sports, or weather are sent via text to the handset; for an additional fee users may upgrade the service to include additional offerings.	
Verizon Wireless	Digital Numeric Paging available for $2.99 per month.	Digital Numeric Messaging allows senders to dispatch their numeric messages via the company's Web site.	Messages cannot exceed 120 characters. The system has the ability to store and resend unsuccessfully delivered messages for up to 14 days.

Table 4.4 Mobile Wireless Text Messaging Plans (continued).

Provider	Price	Services	Limits/Features
Verizon (cont.)	Internet Text Messaging available for \$4.99 to \$9.95 per month depending on location.	Internet Text Messaging enables Web-based and email-based message dispatch.	The service imposes a limit of 10 messages per hour, and offers protection from "spamming."
	Operator Assisted Text Messaging for \$19.95 to \$24.99 per month depending on location.	Operator Assisted Text Messaging offers the convenience of a live operator and includes PC-based dispatch.	
	PC-Based Text Messaging with the basic software costs \$9.99 or an upgraded version in available for \$24.99.	PC-based Text Messaging offers advanced features such as address book/group maintenance.	
Sprint PCS	Two message-specific plans are available to choose from: \$1.99 per month for 10 messages or \$9.99 for 200 messages.	Message dispatch includes web-based, email-based, and PC-based messaging.	Sprint PCS messages have a 100-character maximum and can be stored and forwarded for up to three days.
	Messaging is also bundled within other service plans such as the Wireless Web.	A message tracking page on the Sprint PCS Web site is offered to allow message originators to obtain a tracking number that can be subsequently used to determine if messages have been delivered.	

Table 4.4 Mobile Wireless Text Messaging Plans (continued).

Provider	Price	Services	Limits/Features
Sprint (cont.)		Wireless Web updates from Yahoo! are included for customers enrolled in any Wireless Web Messaging option.	
Nextel	$3.00 per month or $9.00 per month including caller ID.	Message dispatch includes Web-based, email-based, and PC-based.	A Nextel page cannot exceed 280 characters.
		The Web-based dispatch is one of the most flexible available. The message status feature of the dispatch site offers delivery notification. The system allows delivery confirmation to be delivered to an email address or another Nextel phone. Group page allows for dispatch to as many as 20 phones with a single keystroke.	After entering a message, it can be sent immediately or scheduled for delivery at a later date and time. The network has store and forward functionality for up to 5 days.
		NexNote and NexNote Plus allow senders to compose, send, and track the receipt of text messages using a PC and modem. NexNote Plus facilitates the maintenance of lists and dispatches to multiple phones.	The auto callback feature allows users single-key call initiation while reading a numeric message or a phone number embedded within a text message.

line portals, like AOL and Yahoo, have very strong positions and brands. In many respects, these portals have more substantial brands than the carriers, particularly in the Internet space, and subscribers will want to extend their access to these services to include wireless. In Europe and Asia, the carriers have much stronger brands than the wireline portals, and there is less inherent demand for extending wireline portals. Additionally, many of the wireline portals in Europe are owned by the same telecommunications companies that provide wireless services.

Potential Revenues

There are a number of potential sources of revenue that will result from enabling the mobile wireless Internet (see Figures 4.38 and 4.39). First and foremost will be the increase in network traffic. Additional opportunities include wireless ISP services, advertising, m-commerce, messaging, and multimedia. Morgan Stanley estimates that by 2005 these services could amount to $15 billion in revenues in the U.S., and increase ARPU across the entire subscriber base by almost $5 per month, or about 10% of total revenues. Strategy Analytics pegs total cellular data revenues for the U.S. at about $22 billion, or about $8 in additional ARPU across the entire subscriber base (as shown in Figure 4.39). Although there is a large variance between these estimates, the growth rates are consistent. Some of the difference may be due to market definitions and segments included in the analysis. However, there remains a wide disparity between messaging forecasts.

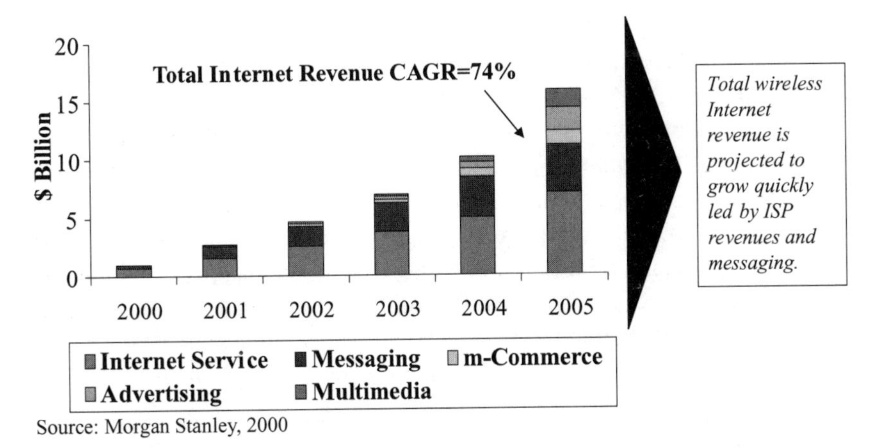

Figure 4.38
U.S. Wireless Internet Revenue Composition.

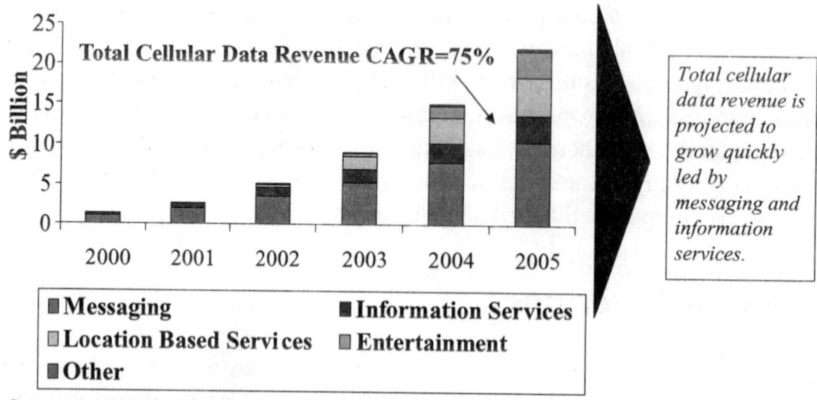

Source: Adapted from Strategy Analytics, Inc.: a Boston-based company (2000b)

Figure 4.39
U.S. Cellular Data Revenue Composition.

Bluetooth

Bluetooth is a royalty-free global wireless technology standard. In early 1998 a group of computer and telecommunications companies that included Intel, IBM, Toshiba, Ericsson, and Nokia began to develop a way for users to connect a wide range of mobile devices without cables. This standard quickly gained new members including 3COM/Palm, Compaq, Dell, Lucent Technologies, and Motorola. Bluetooth membership numbered more than 2000 by year end 2000.

Bluetooth uses tiny radio transceivers that operate in the unlicensed 2.4-GHz band to transmit voice and data at up to 721 kbps. Bluetooth communicates within a 10-meter perimeter and does not require line of sight to establish a connection. According to Intel, the cost of the transceiver is about $20, and should eventually fall to $5.

Bluetooth will enable many new and important wireless data capabilities. For instance, a document could be downloaded to a mobile handset and then printed on a Bluetooth-enabled printer, or a presentation could be carried on a PDA and then shown through a Bluetooth-enabled digital projector. Bluetooth can also be used to automatically synchronize PIM information between devices, when they are brought into range.

The ultimate value of Bluetooth will be determined by the number of Bluetooth-enabled devices available. Bluetooth will adhere to Metcalfe's Law, where the utility of the network is equal to the square of the number of nodes. Another example is the adoption pattern of the fax machine. Fax machines were not very valuable until there were a large number of fax machines with which to exchange faxes.

IDC expects Bluetooth volumes to ramp very quickly, achieving penetration in over 100 million devices per year by 2004 (see Figures 4.40 and 4.41). This is extraordinarily fast growth that may prove to be a bit optimistic, at this point, given the cost and timing of hardware cycles. However, once it becomes clear that there will be substantial actual demand, versus forecast demand, manufacturers will get on board quickly.

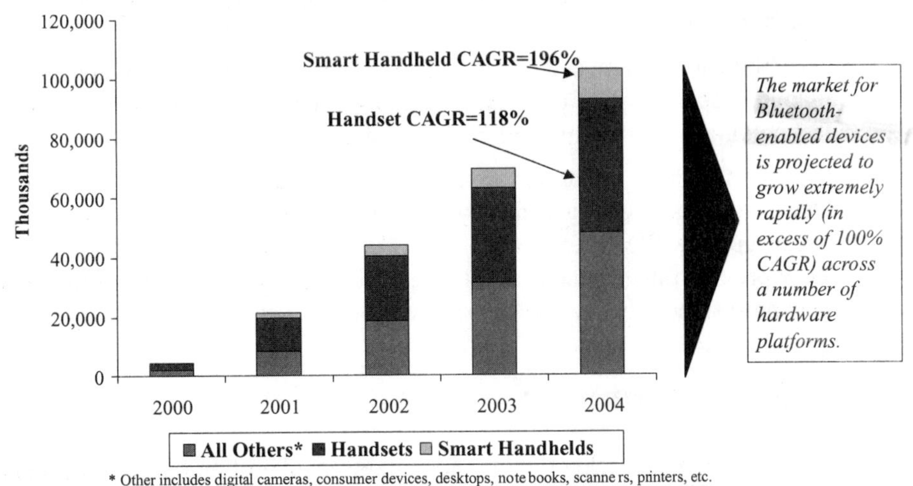

Source: IDC, 2000

Figure 4.40
U.S. Bluetooth Shipments by Hardware Segment.

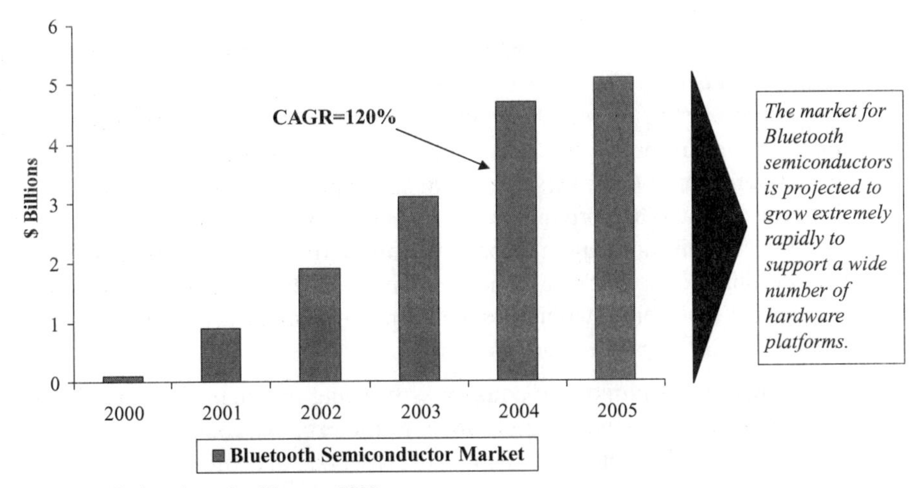

Source: Allied Business Intelligence, 2000

Figure 4.41
Bluetooth Semiconductor Market Worldwide.

Synchronization

One of the key enablers for the mobile wireless Internet will be synchronization. Synchronization is the ability to work off the same master list for appointments, contacts, and email without constantly having to update each device individually. To realize the maximum value from synchronization, updates should take place automatically and in real time.

Synchronization can be local, using wired or wireless technologies. Wired technologies include cables or docking stations, such as for Palm devices, while wireless technologies include Bluetooth, infrared, and 802.11b.

Conversely, synchronization can be enabled using a central database. Fusion One is an example of a company that runs a centralized synchronization database, which updates PIM information on PDAs, handsets, and PCs through the Internet. They provide this service at no charge. This is certainly a more elegant solution, but may pose some security and/or privacy concerns for some individuals or companies.

Voice Recognition

Voice Recognition technology is another key enabler of mobile wireless data/Internet, and will solve many of the input issues surrounding wireless devices. For many years, voice recognition has been predicted to be just over the horizon, but it has proven to be a particularly thorny problem. The software has never seemed to be able to handle the wide range of vocabularies and accents well enough to be able to launch a product with wide acceptance.

If a commercially acceptable product can be developed, the potential for speech recognition is vast. The applications for speech recognition technology also extend far beyond the realm of mobile wireless communication, and are limited only by the imagination. In mobile wireless communications, though, speech recognition would be a boon. This would solve many of the input issues surrounding the keypad and increase safety. When driving a vehicle, operating a wireless handset is often very distracting, and is drawing increased attention from lawmakers. Speech recognition would allay some of these concerns. This will be particularly true for telematics, which may be limited without speech input. Telematics is an integrated mobile telephony and GPS system in an automobile.

There are a number of companies working on voice recognition, and again the solution is just over the horizon. Some of the primary companies working on speech recognition are Lernout & Hauspie, Speechworks, and IBM.

Location-Based Services

Location-based capabilities will be one of the most important functions and differentiators for mobile wireless data/Internet services. The ability to provide timely, pertinent, and location-specific information to a subscriber is tremendously valuable, both to the content provider and to the subscriber. From a marketing perspective it is tremendously valuable to know the identity of an individual, the location of an individual, and what the individuals most interested in at that particular moment.

As a result of location-identifying technology, a myriad of revenue-generating, location-based services can be made available. Some of the major classes of applications can be tracking, guiding, and notification. These services become even more compelling in a packet-based environment, where information can be proactively pushed to the recipient, instead of having to be speculatively retrieved.

Tracking applications would be valuable in locating and tracking assets, like fleets of vehicles. Other potential uses include ascertaining the location of children, employees, pets, or even criminals.

Guiding applications could assist customers in finding specific locations and addresses. These services could provide step-by-step directions to a particular destination, and also be used to find the nearest Chinese restaurant, ATM, or shoe repair store. This application is considered to be a very attractive advertising and revenue-generating opportunity.

Notification applications can be used to alert subscribers regarding stock trades, traffic situations, flight delays, and email. This application is particularly useful when enabled by a packet-switched network where information can be pushed to the subscriber in a timely fashion.

Location-based services are expected to grow very rapidly and generate substantial revenues. Maintaining privacy is a key concern with making location information known. People are concerned that the government, employers, or businesses will have too much ability to track their movements.

Please see a further discussion of location services in Chapter 3.

Devices

There will be a wide array of devices that will access mobile wireless data and Internet services. The main classes of device that will be used to access these services will be PCs, wireless handsets, PDAs, and pagers. The availability of powerful, reasonably priced devices will be of paramount importance to the uptake of data services. This fact is cited by network operators who are planning to roll out GPRS services. In late

2000, European carriers found that they were not able to secure enough GPRS handsets to adequately test their networks and services let alone launch services for public consumption. The handsets that they were able to secure operated at speeds well below the advertised 115 kbps. This was due to excessive heat generation observed at relatively low data-transfer rates. Additionally, achieving higher throughput requires combining time slots with a resulting loss in voice capacity. Thus, the carrier has to calculate the best balance of data and voice capacity for its network.

Terminal availability and capability are likely to be one of the key limiting factors in the early days of wireless data services. This is currently a factor with GPRS, as widespread availability is not expected before the second half of 2001, at the earliest. Motorola was the only manufacturer shipping GPRS-enabled handsets in late 2000. This situation is likely to persist with 3G networks as well.

PCs are able to access wireless data services by using wirelessly enabled PCMCIA cards and by using wireless handsets as a modem.

Structural Market Assessment

The market structure for wireless data network carriers is likely to be very competitive (see Figure 4.42). This is forecast to be an extremely fast-growing market that will drive tremendous increases in network traffic. However, participating in the market

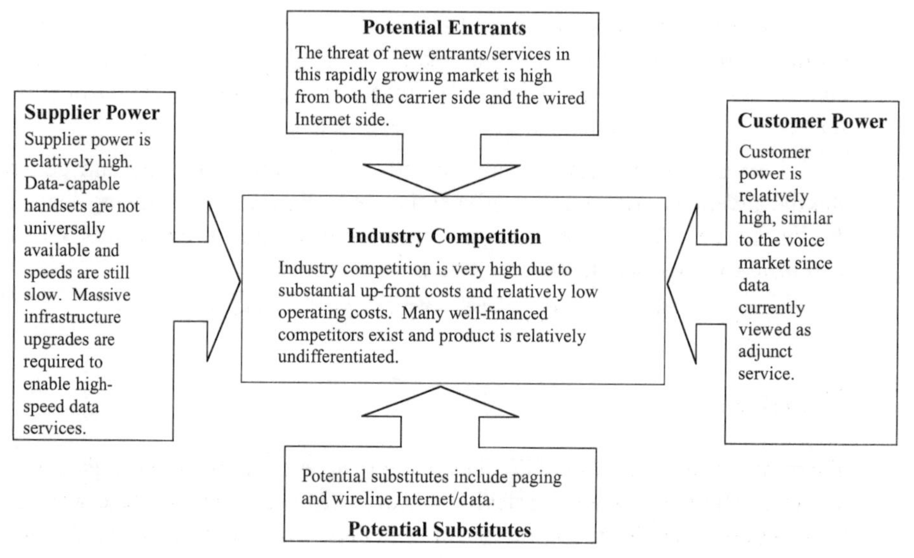

Figure 4.42
Structural Market Assessment: Cellular Data Network Service Providers.

requires substantial investment in infrastructure and support. Additionally, content does not appear to have a high degree of differentiation between carriers, and it is uncertain how much consumers are willing to pay for these services.

Customer Power

Customer power is reasonably high, for the same reasons that it is in the voice market. Data is also regarded as an adjunct service, and not the primary reason for securing service. Thus, the customer power in the voice market directly impacts the data market, particularly if data services are undifferentiated.

Conversely, a very well-crafted and differentiated data offering may serve to counteract customer power in the voice market, creating a huge win for carriers.

Potential Substitutes

Potential substitutes include the traditional landline Internet and paging/data networks. However, the compelling benefits of mobility will drive rapid growth in mobile wireless data. Additionally, cellular networks will have vastly greater capacity than the paging data networks. This will allow the cellular data networks to offer much more interesting and compelling content. For the more basic content, cellular network operators will be able to bundle content with other services much more economically than paging/data network operators, giving the cellular operators a substantially better value proposition.

Supplier Power

Supplier power is relatively high due to the lack of sufficient quantities of data-capable handsets and the need for massive, complicated infrastructure upgrades. This is likely to persist for the foreseeable future.

Potential Entrants

The threat of new entrants is very high both in terms of carriers substantially expanding their offerings and other data content suppliers, such as an AOL or Yahoo, trying to displace the carriers. Carriers can be expected to massively overhaul their networks to enable new and exciting data services.

Industry Competition

Industry competition can be expected to be very high since there are a number of well-financed competitors counting on this market to drive their future growth. Additionally, there is not much differentiation apparent in the competitors' data offerings, and the most popular services are apt to be included in the basic service charge.

Outlook

Mobile wireless data/Internet services represent a tremendous opportunity for both subscribers and providers. Subscribers will gain access to vital data from anywhere and at any time. This will substantially increase their capabilities in both their personal and professional lives. Carriers will have the opportunity to drive substantial incremental traffic over their networks, charge for new value-added services, and reduce churn.

The outlook for mobile wireless data/Internet is very favorable from both a growth and a penetration standpoint. However, a number of key hurdles remain before high-speed packet-switched Internet and data services are universally available. These challenges include addressing the variety of interface issues, the availability of data-enabled handsets, and the deployment of high-speed networks. These issues may retard or delay the growth of mobile data/Internet services, proving forecasts in 2000 to be overly optimistic.

In terms of revenue growth, there is substantial opportunity to increase revenues with new, innovative, value-added services. Mobile wireless data/Internet is likely to become a multibillion dollar industry in the not-too-distant future. However, carriers will have to be clever in designing new services, as they are unlikely to be able to extract a high price for messaging and basic information. These types of services are apt to be included in every basic package. Carriers will also need to be cognizant that prices for the transmission of both voice and data are likely to fall quickly, and need to be offset by rapid gains in utilization.

Competition should be fierce in mobile wireless data/Internet, and not only between carriers, but between traditional wireline content providers, such as Yahoo and AOL, as well. Content providers will seek to displace the carriers, as they have done in the wired world, relegating them to the role of dumb pipes. Carriers can be expected to put up a significant fight in this battle and leverage all of their inherent advantages, such as control of the handset, control of the network, control of location information, and control of the customer. However, when AOL is shopping for a terabyte per month of transmission capacity, they are likely to find many interested suppliers, given the economics of running a wireless network. This is particularly true if any of the major players starts falling behind, as the amount of traffic generated by a partner like AOL could lead to an immediate reversal of fortunes.

In the short term, the biggest beneficiaries of wireless mobile data/Internet are likely to be the infrastructure suppliers, since massive system upgrades are necessary to enable mobile data. These suppliers include the normal wireless infrastructure suppliers like Ericsson, Motorola, Nortel, Nokia, and Lucent, but also the data gateway/server suppliers like Openwave Systems. As is the case in the voice world, infrastructure suppliers are likely to be in the most attractive position from an industry attractiveness perspective.

All in all, mobile wireless data/Internet is a huge positive for the industry and will drive many tremendous innovations. However, over the medium term, it is likely that forecasts of uptake and incremental revenues are on the optimistic side. Additionally, there are huge investments required to upgrade the infrastructure and terminals, and there is fierce competition between a relatively large number of carriers. This situation calls into question the prospect for meaningful returns on invested capital for the carriers.

PAGING/DATA CARRIERS

Paging/data carriers wirelessly communicate data to their subscribers. The paging market has grown very rapidly since its inception in 1956, to serve over 40 million subscribers. Paging data can be tone, numeric, or alphanumeric.

The first paging systems alerted customers to an incoming message through a tone. The subscriber then called in to an operator to retrieve the message. Numeric paging was then developed, and alerted the subscriber through a number that appeared on a small LCD display on the pager. This was generally a phone number. Alphanumeric paging allows the subscriber to receive text information on displays of various sizes.

Traditionally, paging has been a one-way medium. However, with the advent of ReFLEX, Mobitex, and ARDIS, interactive mobile data communication has been enabled (see Table 4.5). This allows customers to send and receive email, using a QWERTY keyboard, with their pager.

However, the paging market may have seen its high-water mark. Mobile wireless telephony is proving to be a very substantial threat to the paging market, and for the first time, paging subscribers declined in 2000. The key advantages of paging versus mobile wireless telephony are superior in-building penetration, wider coverage areas, and longer battery life.

Table 4.5 Paging Network Comparison.

	Description	Applications	Frequency	Data Speed	TCP/IP
POCSAG	Low speed 1-way	1-way numeric and alpha	Any paging frequency	512–2400 bps	No
DataTac	Proprietary data messaging from Motorola	2-way data messaging	800 MHz	Up to 19.2 kbps	Yes
FLEX	High speed 1-way	1-way numeric and alpha	Any paging frequency	6.2 kbps	No
ReFLEX 25	2-way messaging and data protocol	2-way short message	Out 929–932, 940–941; in 896–902 MHz	6.4 kbps	No
Mobitex	Proprietary data messaging from Ericsson	2-way messaging data	900 Mhz	Up to 19.2 kbps	Yes

Source: Redman 2000

Paging Market Size and Forecast

Paging is an enormous market in terms of number of subscribers. For 1999, IDC estimated that there were 41.5 million paging subscribers generating $4.3 billion in revenue (see Figure 4.43). This implies annual revenue per unit of $104 or monthly ARPU of about $8.67.

IDC projects that the paging market will continue to grow, albeit at a slow rate (3.9% subscribers, 5.6% revenues) for the next several years. However, this may prove to be an optimistic forecast, given subsequent industry reports and commentary.

Arch, by far the largest paging company after its merger with PageNet, reported in its September 30, 2000, 10-Q, that "Arch expects revenue to continue to be adversely affected in 2000 and 2001 by declining demand for basic numeric and alphanumeric paging services...and believes that it will experience a net decline in the number of units in service for 2000 and 2001... Arch's addition of two-way messaging subscribers is likely to be exceeded by its loss of basic paging subscribers." They also indicated that they felt there had been no growth in the industry in 1999.

Arch reported losing a net 571,000 subscribers, or 8.2% of its then total, in the first nine months of 2000. They also reported that PageNet had gone from 10,604,000 subscribers on June 30, 1998, to 7,858,000 subscribers on June 30, 2000, a loss of 2,746,000 subscribers (26%) in just two years.

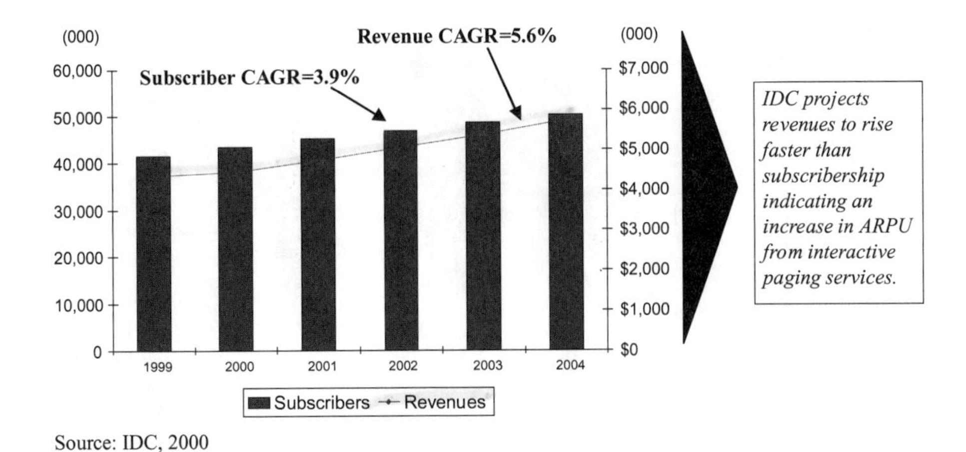

Source: IDC, 2000

Figure 4.43
U.S. Paging Revenues and Subscribers.

The losses in subscribers are attributed to the relatively more attractive value proposition offered by mobile wireless telephony carriers. The cellular carriers often include paging-type services at no incremental cost, on a single device. The wireless handset is also usually subsidized by the carrier.

Paging Market Share

In the face of very challenging financial conditions, the paging market has consolidated. In 2000, the industry was dominated by the top four carriers, and further consolidation can be expected (see Figure 4.44).

Arch/PageNet, which is the product of a merger between the two largest paging companies, controls more than 40% of the market. This merger was completed in November 2000, after a Chapter 11 reorganization by PageNet. In June 1999, Arch acquired MobileMedia, the fourth largest paging company at the time, after MobileMedia filed for bankruptcy.

The second largest paging company in 2000 was Metrocall. Combined, Arch/PageNet and Metrocall controlled about two-thirds of the market.

BellSouth Wireless Data (BSWD) and Motient are small players in the overall paging market. However, their unique network capabilities (packet switching) offer them real opportunities to grow. Both BSWD and Motient offer service for the RIM Blackberry on their networks and BSWD is the sole provider of the Palm.net service, the proprietary wireless ISP for the Palm VII.

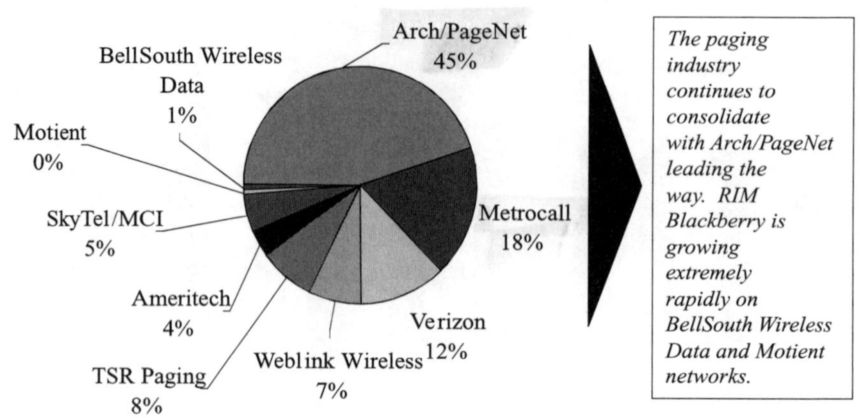

Statistical data from The North American Wireless Marketplace 2 nd Quarter 2000 Update February 5, 2001, by Paul Dittner and Bryan Prohm, Gartner

Figure 4.44
U.S. 2000 Data/Paging Subscriber Share by Carrier.

Customer Base

The vast majority of paging subscribers—more than 80%—are business users (see Figure 4.45). Often these are emergency personnel, salesmen, service personnel, and support staff. These individuals continue to need the superior coverage and in-building reception that paging offers. This is in contrast to mobile phones, which are not as capable in these respects, particularly for in-building reception. Therefore, paging will remain desirable to many people, despite the increased capabilities of wireless phones.

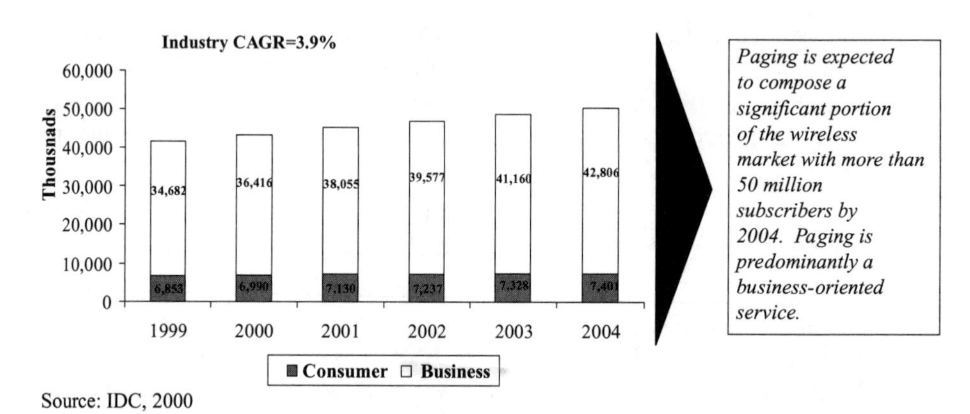

Source: IDC, 2000

Figure 4.45
U.S. Consumer and Business Paging Subscribers.

However, the industry will continue to be plagued by the threat of mobile wireless telephony. This exerts significant downward pricing pressure on the industry even if the subscriber does not replace paging altogether.

A quote from Arch's September 2000 10-Q illustrates the impact mobile phones have had on pagers: "mobile wireless phone services now include wireless messaging as an adjunct service...it is less expensive for an end user to enhance a cellular, PCS, or other mobile phone with modest data capability than to use both a mobile phone and a pager. This is because the nationwide carriers have subsidized the purchase of mobile phones and because prices for mobile communications have been declining rapidly. In addition, the availability of coverage for these services has increased, making the two types of service and product offerings more comparable."

This situation highlights the impact on industry attractiveness of "potential substitutes." Paging carriers are going to have to develop greater differentiation in their product and its applications in order to survive. This will be particularly true for their main user base, business people. Businesses will be unwilling to pay for duplicate services and business people are loath to carry any superfluous device.

PageNet has supplied concrete evidence of the difficult pricing environment. In 1999, PageNet attempted to raise prices on its paging services. This resulted in an increase in ARPU of $.09 per month or 1.2% ($7.71 to $7.80). However, total revenue went down by 5.1% as they lost 11.1% of their subscribers. Although not all of this decline should be attributed to pricing, the evidence clearly indicates a high degree of price elasticity. The price elasticity in this instance appears to be much greater than 1. This example further confirms the tremendous price elasticity for wireless communications. However, in the past prices have generally fallen and driven larger percentage increases in revenues; this case highlights the two-edged nature of price elasticity and the impact of increasing prices.

The paging industry, despite its predominantly business constituency, suffers from very high churn levels. Weblink Wireless, formerly known as PageMart, reported monthly churn levels of 2.5%, 3.2%, and 3.1%, for 1997, 1998, and 1999, respectively. A 3% monthly churn level translates into a 43% annual churn rate. This is an enormous number of customers to be disconnecting any type of service, and is particularly alarming in a service sold to business people who face reasonably high switching costs, like updating business cards and stationery.

Interactive Paging

Interactive paging represents a primary example of how paging carriers have tried to improve and differentiate their service. Interactive, or two-way, paging is the only segment of the market that is actually growing. This has been exemplified by the tremendous growth of the Motorola two-way pagers and the Research In Motion (RIM) BlackBerry pagers.

The classes of interactive service can be defined by their level of interactive capability. These classes include 1.5-way pagers, where the device sends a receipt or confirmation back over network; 1.7-way pagers, where the device has two-way communication with optional preset responses; and full two-way pagers that transmit original messages and responses with a QWERTY keyboard.

Two-way paging is expected to grow at the expense of, or cannibalize, one-way paging (see Figures 4.46 and 4.47). While not growing the overall market, two-way paging may prevent the industry from going into complete freefall, as one-way paging provides a much more murky value proposition for the customer. However, it is important to note that, according to Arch, two-way is not adding net new customers to the industry, since one-way customers are being lost more rapidly. This is in conflict with the IDC projection of net industry growth in subscribership. Strategy Analytics also foresees the rapid ascension of interactive paging and decline of numeric paging. Their revenue forecast anticipates no industry growth over the intermediate horizon (goes up then declines).

Structural Market Assessment

The market structure for paging service providers is currently very unfavorable (see Figure 4.48). The industry is very unprofitable and highly competitive. Substantial investment is required to enable two-way paging and growth is slow to nonexistent, perhaps even negative. The major firms in the industry are under serious financial duress, and many may go into bankruptcy.

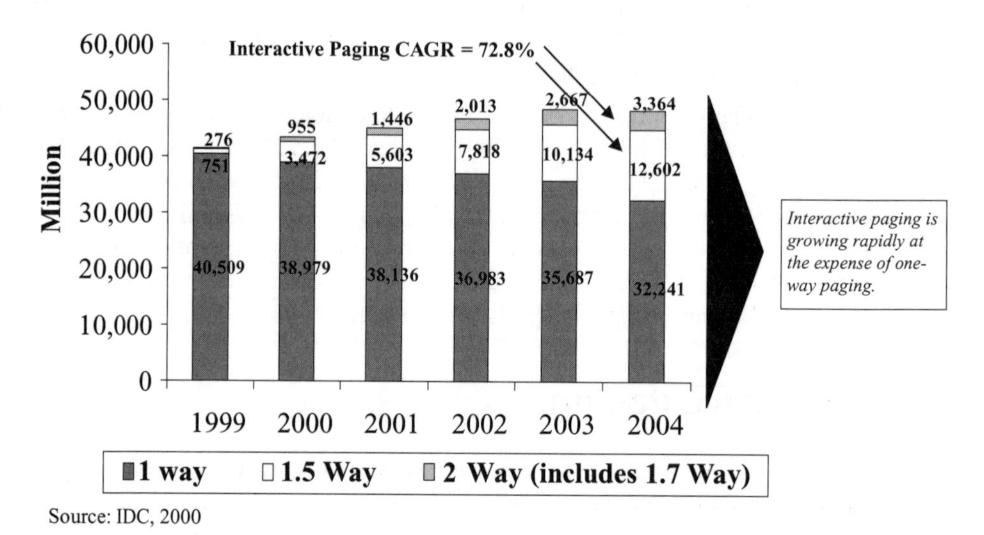

Figure 4.46
U.S. Paging Subscribers by Service Type.

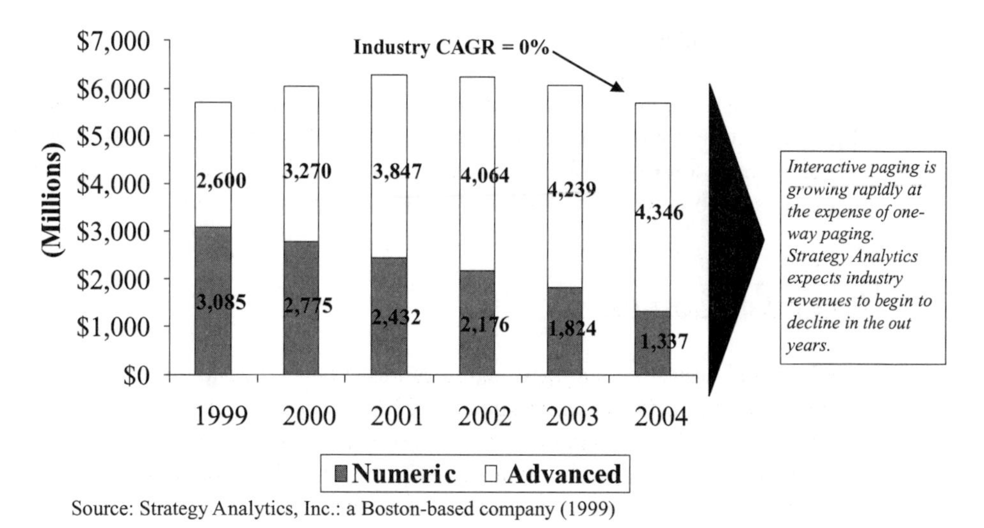

Source: Strategy Analytics, Inc.: a Boston-based company (1999)

Figure 4.47
U.S. Paging Revenues by Service Type.

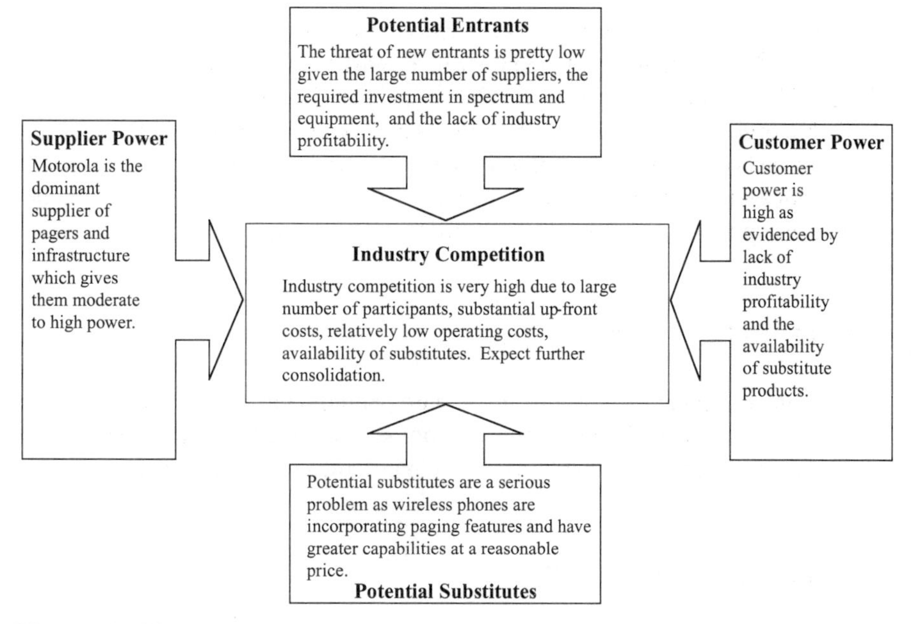

Figure 4.48
Structural Market Assessment: Paging Network Service Providers.

Customer Power

Customer power is high as evidenced by the lack of profitability and high levels of churn. The industry is continuing to consolidate; however, there are still ample competitors for customers to choose from. There are also many substitutes, particularly wireless telephony, which is leading to a net decline in subscribers.

Potential Substitutes

The key substitute for paging is wireless telephony. The situation is exacerbated as wireless handsets add voicemail, data, and SMS messaging capabilities. Additionally, as 2.5G and 3G networks are rolled out, the bandwidth of these networks will swamp the capabilities of the paging/data networks. Paging/data networks will be unable to respond convincingly to this challenge, given their spectrum constraints.

Supplier Power

Supplier power is moderate to high. Motorola—with their FLEX system—has been the dominant supplier to the industry with more than 90% market share. Motorola has exited the paging infrastructure business and has licensed their technology to Glenayre Technologies. This leads to substantial supplier power in any negotiations, and carriers are beholden to a potentially limited technology roadmap. However, supplier pricing power is moderated by the difficult conditions in the industry, which may also limit a supplier's investment in creating new services.

Potential Entrants

The threat of potential entrants in the pure paging/data carrier market is low, given the high up-front costs, large number of competitors, and lack of profitability. However, the threat of potential substitutes is high, particularly from the wireless telephony side.

Industry Competition

Industry competition is fierce with a large number of competitors facing a very difficult environment. The economics of paging encourage substantial price competition with a commodity product, high sunk costs, and relatively low operating costs. The industry has been consolidating rapidly and this should be expected to continue. Many of the same economic issues that face the telephony carriers impact the paging operators, such as high acquisition costs, high churn, and declining pricing. In addition, there is the widespread availability of an attractively priced and featured substitute.

The paging industry may provide an analog for the future of mobile wireless telephony. The consolidation and challenging financial conditions sweeping through the paging industry are likely to be mirrored on the telephony side, particularly if consolidation is restricted.

Outlook

Although the paging industry may be struggling mightily and very much out of favor today, writing the industry's obituary may be premature. Paging has some inherent advantages in coverage, in-building penetration, and extended battery life. Additionally, paging companies own spectrum, which, although limited, should not be discounted. Spectrum is like oceanfront property, and rarely goes down in value, especially in the long term.

Furthermore, paging companies can substantially expand their applications into telemetry. Telemetry is machine-to-machine or machine-to-human communication. The surface has barely been scratched in this market and paging's capabilities make a good fit. For instance, monitoring in-building machines, like copiers or vending machines, could be valuable. Monitoring applications require two-way communication, but generally need very little bandwidth.

Therefore, with a change in focus, the paging/data network operators may be able to find a defensible market consistent with their strengths. Much of this will depend on the paging operators being able to survive financially.

5 Network Service Consumers

In this chapter...

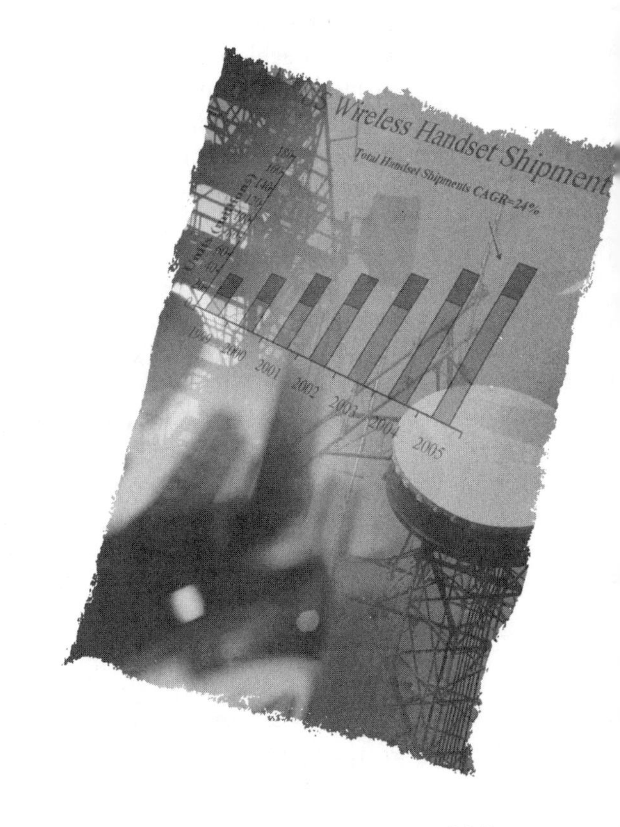

There are three distinct classes of consumer for mobile wireless communications services—retail, business, and value-added service provider. The three customer segments are growing rapidly; however, they differ significantly in terms of what motivates their purchases.

Retail consumers are individual subscribers who use wireless services primarily for communication, convenience, and safety. Business consumers have two primary motivations for using mobile wireless communications—to increase the productivity of their employees and/or to better address the needs of their customers. Value-added service providers seek to use somebody else's network to provide their own service. Examples of value-added service providers include Wireless Internet Service Providers (WISPs), Wireless Application Service Providers (WASPs), telematics (a combination of wireless communications and global positioning in an automobile), and telemetry (machine-to-machine or machine-to-human communication).

Revenues and subscribers in the mobile wireless communications industry have compounded at extraordinary rates and this is projected to continue for the foreseeable future (see Figure 5.1). This growth has been fueled by falling prices and improved service. The improving value proposition has attracted many customers to the industry and provided the ability to use the service more effectively. Wireless communications has achieved the price, utility, and service levels necessary to attract both the business market and the mass consumer market. The growth has, in turn, enabled firms to build value-added services targeted at these customers. Additionally, mobile wireless communications growth is inextricably tied to and driven by the three primary growth themes of the U.S. economy—digitalization, the speed and power of information processing, and mobility.

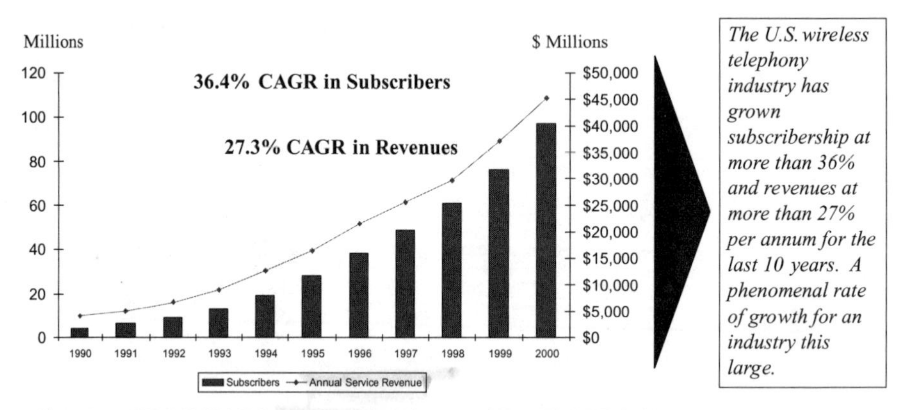

Figure 5.1
U.S. Mobile Wireless Telephone Subscribers and Revenues 1990–2000.

RETAIL CONSUMERS

Retail consumers make up the majority of subscribers to mobile wireless communications; 65–75% of customers can be classified as retail consumers (see Figures 5.2 and 5.3). However, retail consumers spend roughly 50% less than business consumers on wireless services (see Figure 5.4). The retail consumer generally acquires service directly from the carrier (see Figure 5.5) and price is one of the main factors in the retail consumer's decision to obtain service. It is also the primary driver in the choice of handset and provider (see Figures 5.6 and 5.7).

The primary motivations for consumers to obtain mobile wireless communications services are convenience and safety (see Figure 5.8). The increase in personal capability enabled by mobile wireless communications is viewed as very convenient by retail consumers. Also, many subscribers activate service for safety reasons, keeping their wireless handset in their car's glovebox in case of emergency.

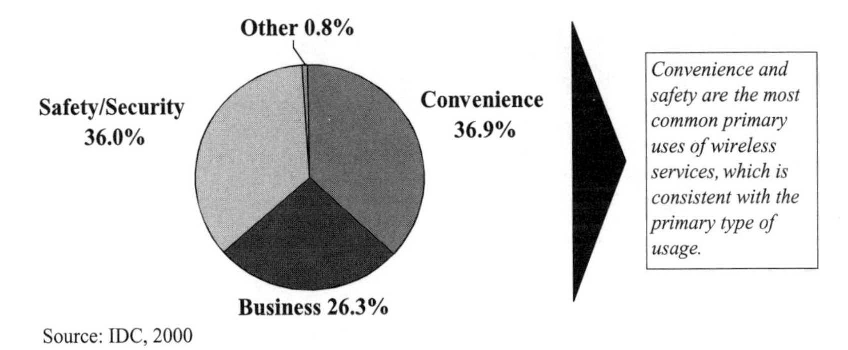

Figure 5.2
U.S. Wireless Subscribers by Primary Use.

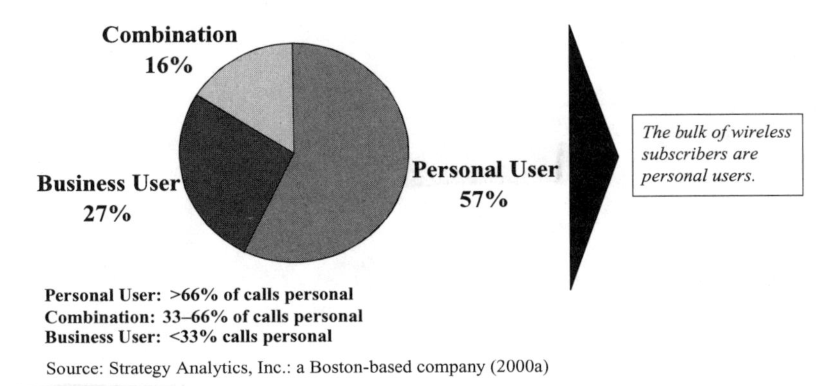

Figure 5.3
U.S. Wireless Subscribers by Type of Usage.

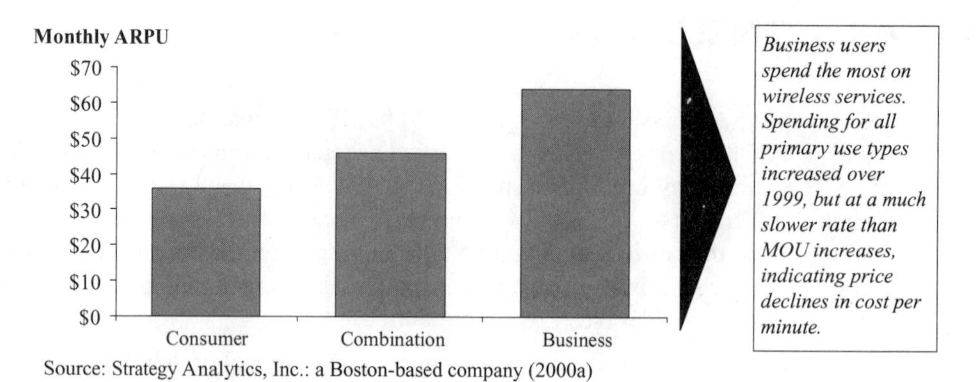

Source: Strategy Analytics, Inc.: a Boston-based company (2000a)

Figure 5.4
U.S. Monthly ARPU by Type of Usage.

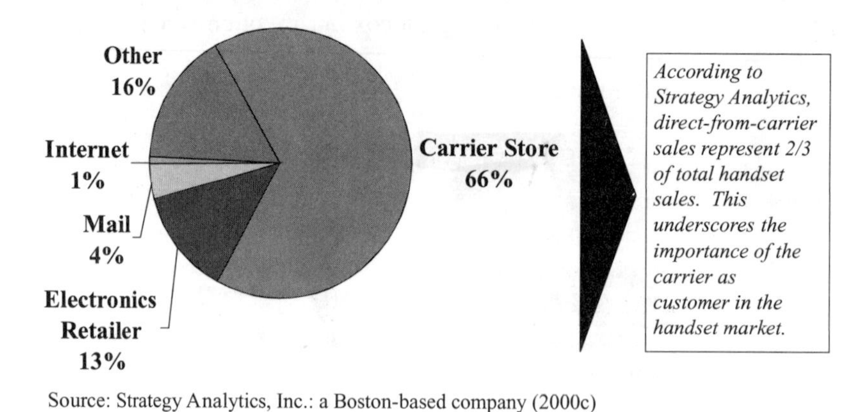

Source: Strategy Analytics, Inc.: a Boston-based company (2000c)

Figure 5.5
U.S. Handset Distribution.

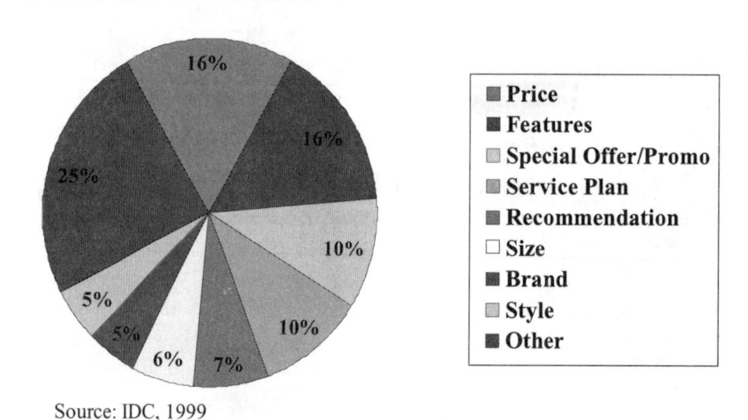

Source: IDC, 1999

Figure 5.6
Factors Influencing Handset Purchase.

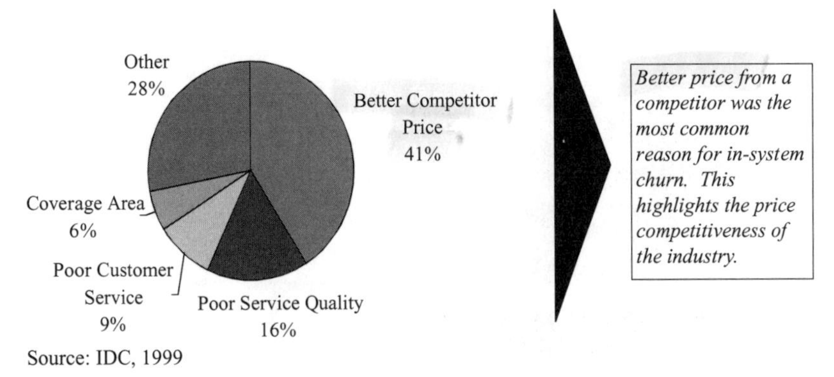

Figure 5.7
Primary Reasons for In-System Churn.

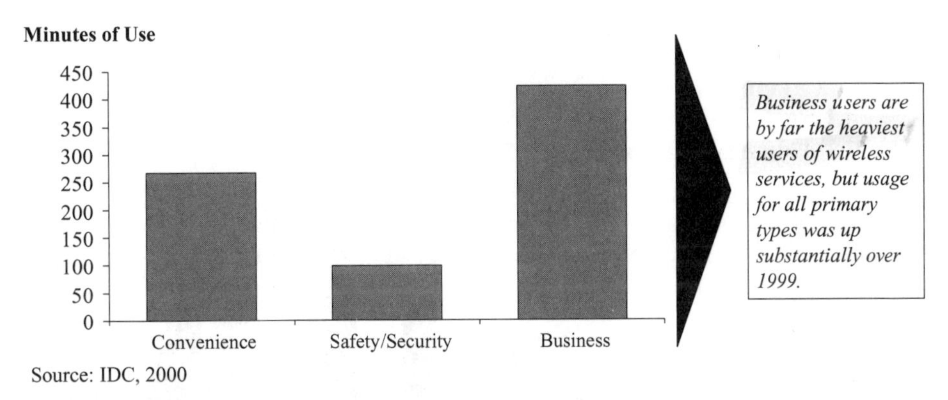

Figure 5.8
Minutes of Use per Month by Customers' Primary Use.

Since retail customers are very price sensitive—both in their service and handset decisions—churn is high and loyalty is low. This is due to the rapid decline in the price of mobile wireless services and the commodity nature of the product. Additionally, retail consumers are eager to upgrade their handsets and use churning as a low-cost method of upgrading their handsets. Switching service providers, in the U.S., requires a new handset, and handsets are substantially subsidized by carriers as part of their marketing budgets. This makes it attractive for consumers desiring a new handset to churn between carriers

As carriers target younger and less affluent consumers, prepaid plans are rapidly gaining in popularity (see Figures 5.9 and 5.10). This may be because many of these younger and more financially constrained customers have not established strong credit and want to limit their spending. This is particularly true when parents are paying for their children's service. According to Salomon Smith Barney, the fastest growth

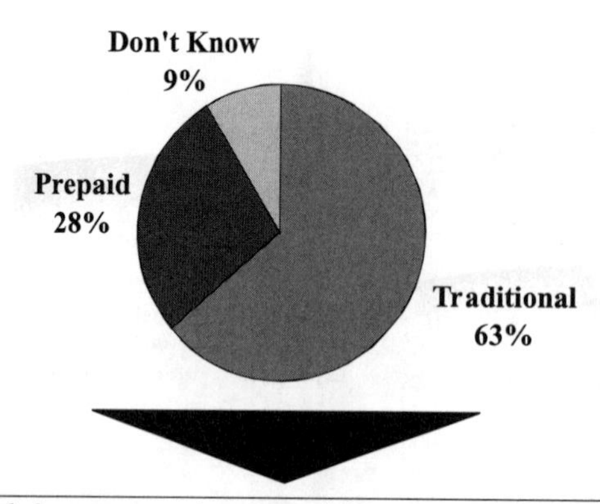

Prepaid plans are a popular means of obtaining wireless service for the credit impaired (approximately 30% of applicants are declined for traditional plans due to credit issues), and those who want to regulate spending (teens, parents, businesses). Prepaid plans are expected to grow rapidly in popularity.

Source: IDC, 1999

Figure 5.9
U.S. Household Subscription Share by Type.

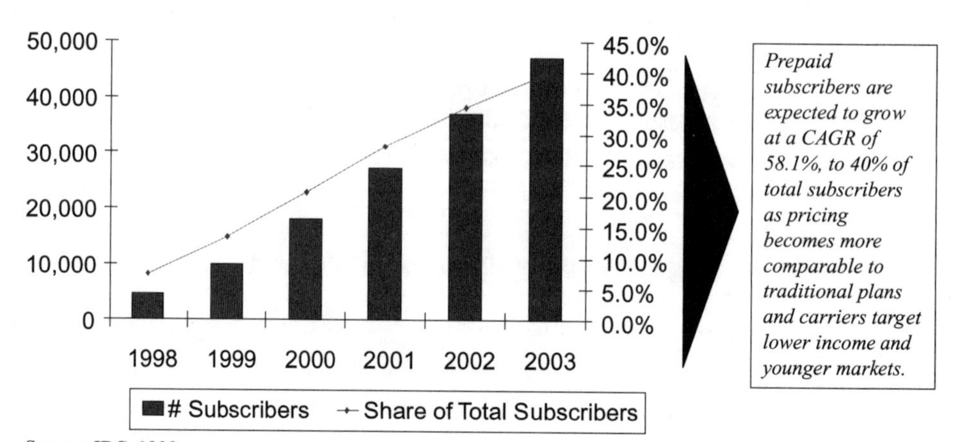

Source: IDC, 1999

Figure 5.10
U.S. Business and Consumer Prepaid Wireless Subscribers.

areas of the market will be in the younger and older ends of the age spectrum. Mobile wireless communications has already achieved high levels of penetration in the 25–54 age bracket (estimated to be 60% in 2000; see Figure 5.11).

Although voice communication is the primary reason consumers subscribe to mobile wireless services, retail consumer mobile wireless data will be driven by

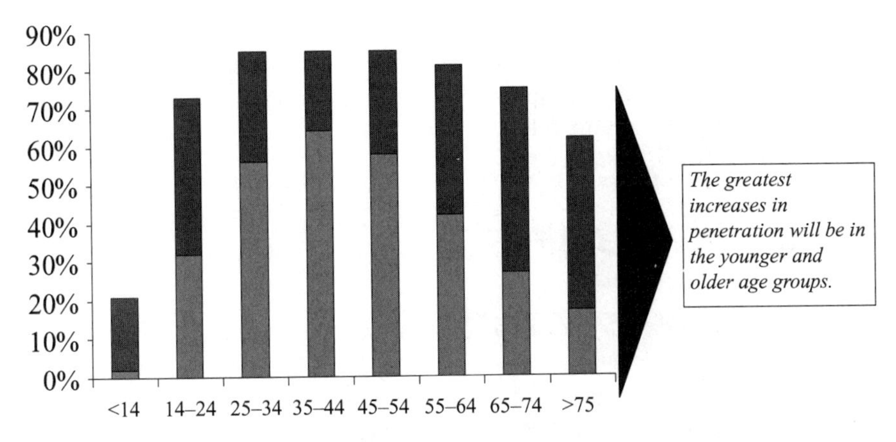

Estimated 2000 Penetration **Estimated 2007 Penetration**

Source: Salomon Smith Barney, 2000

Figure 5.11
Estimated Mobile Wireless Penetration by Age Group.

convenience, timely access to valuable information, and enhanced communication capability, through messaging and email. Email and messaging are expected to be the dominant consumer applications on the wireless Internet, extending the popularity of these applications from the wired Internet. Additionally, location-based content, such as receiving detailed driving directions or local event information, is likely to be a valuable and popular service to many consumers.

As mobile commerce (m-commerce) becomes more widespread, consumers will be better able to make purchases, at their convenience, especially last-minute or urgent purchases. In addition, comparison shopping will flourish as consumers will be armed with product pricing from a variety of retailers, as they visit local merchants. This will be enabled by Universal Product Code (UPC) scanners, which can comparison shop for the particular piece of merchandise across the entire Internet.

Another important service will be financial applications, such as performing stock transactions or banking. These features will be desired by consumers who will find it quicker and more convenient to use the mobile Internet than to deal with a live operator or a voice response system.

Outlook

Retail consumers will continue to be motivated by communication, convenience, and safety. As prices come down and prepaid plans become more widespread, more consumers will be able to subscribe to mobile wireless communications. This will continue to fuel the long-term growth of the market. The addition of wireless data,

Internet access, and location-specific information will lead to vastly increased capabilities for individual subscribers and drive increased penetration and consumption of mobile wireless communications. The consumer market should be expected to grow rapidly for the foreseeable future.

BUSINESS CONSUMERS

The business market, for both voice and data services, is also expected to grow dramatically. Businesses have been adopting, and will continue to adopt, mobile wireless services to improve the productivity of their employees and to improve their customer service. Business users are highly valued by carriers since they spend more on mobile wireless services and churn less frequently than retail consumers.

Going forward, the main focus of mobile wireless services in the business environment will be the integration of voice and data communications to increase the productivity and efficiency of workers. Voice communication is still the "killer app" in mobile wireless communications, but data will become increasingly important as bandwidth and terminal constraints are solved and as more secure applications are introduced.

Email is likely to become the first application to experience widespread adoption by the business community (see Figures 5.12 and 5.13). Email has become the primary communication medium for many corporations and is the most popular application on the wired Internet. Thus, when employees are away from their desks, they are out of touch with their colleagues and customers and are unable to receive

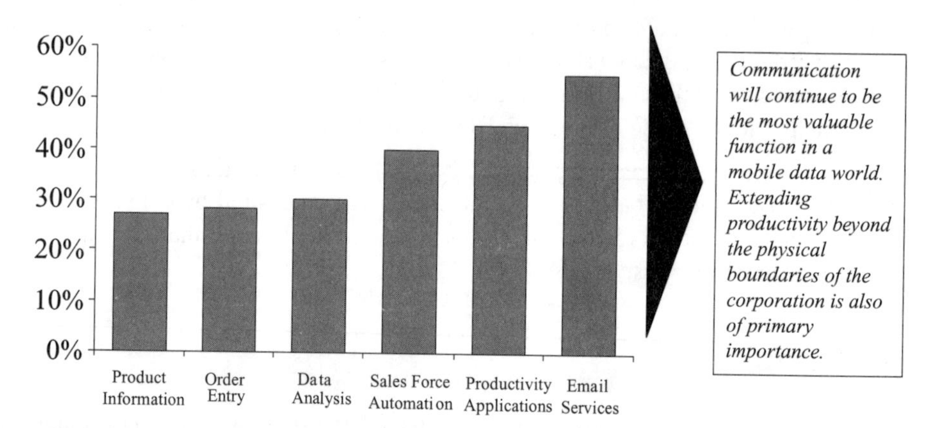

Source: Merrill Lynch, 2000

Figure 5.12
Priority of Mobile Wireless Data Services for Corporate Customers.

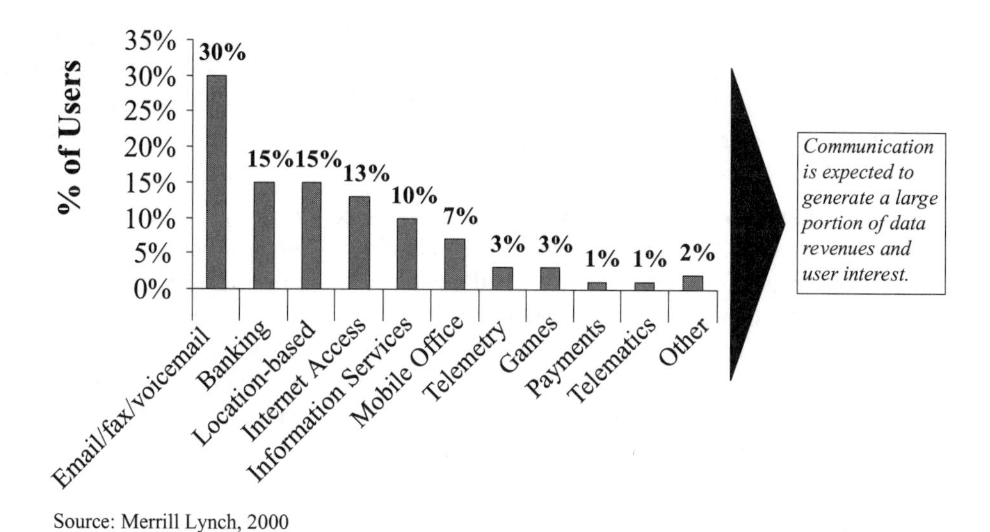

Source: Merrill Lynch, 2000

Figure 5.13
GPRS Mobile Wireless Data Revenues by Application

information and respond to requests in a timely manner. Wireless access to email will significantly increase productivity and effectiveness by allowing workers to be in constant communication with their associates. Access to calendar functions and contact lists will also enhance worker productivity and efficiency.

In the future, many mobile employees will be reachable by dialing a single phone number, which will ring as needed, on their wireless or fixed-line phone. This is a service that has already been implemented in the USWEST/Qwest service area. Employing a universal phone number is a large step in the direction of constant connectivity; it will substantially increase responsiveness and productivity by eliminating phone tag and the leaving of a single message in multiple mailboxes.

Over time, many more applications will be wirelessly enabled. This will increase efficiency by giving employees the ability to place orders, track shipments, or access necessary company information from any location. For example, a salesperson will be able to check product inventory and place an order with a wireless device in real time. Thus, that salesperson can make more sales calls and meet customer requirements more effectively. However, corporations will not subscribe to mobile data/Internet products until they are reliable, fast, and secure. The first applications adopted by corporations are likely to be simple and limited extensions of existing wired services and focus on sales force automation and productivity applications.

Wireless devices will offer businesses a new medium of customer interaction that will afford greater customer intimacy and reduced costs. This parallels the fixed-line Internet, which was used to improve customer satisfaction and reduce costs. The use of wireless data devices for customer service and interaction will substantially reduce operating costs by avoiding the use of live customer representatives and by reducing mistakes. The reduction of ordering mistakes has been found to be one of the most valuable features of online ordering for the wired Internet. Customers will also be able to exert more control over the interaction by receiving service through a wireless handset from any location.

Some examples of wireless data devices being used as a tool to increase customer satisfaction could include disseminating time-sensitive information such as hotel or car rental information when customers land at a local airport, and allowing customers to access information, goods, or services at their convenience via the wireless device. For business-to-business transactions, wireless data devices may be given to important clients to track shipments, order goods and services, and receive the latest product information. This has the potential to substantially increase customer switching costs and to promote increased customer loyalty.

Wireless devices also have the potential to become universal payment devices through their unique equipment identifier. Wireless data devices will be able to make convenience purchases at vending machines equipped with wireless transceivers. This application has the potential to decrease the loss of sales as a result of customers not having change or small bills. This application is also thought to be a good fit for parking meters.

Wireless devices will also be able to make payments for goods and services purchased over the Internet. With payments being made from a wireless data device, human interaction and mistakes are minimized, further reducing operating expenses. The savings can be truly compelling (see Figure 5.14).

Further enabling the use of wireless devices as a payment mechanism was passage of the electronic signature bill into law. This law recognizes online signatures as

Channel	Cost (US$)
Branch	1.07
Telephone	.54
Voice Recognition	.27
PC	.26
Fixed Internet	.13
Mobile Internet*	.16

The Internet (either mobile or fixed) offers tremendous cost savings for customer service. This will be a compelling reason for businesses to encourage wireless access for their customers.

Source: Booz, Allen & Hamilton, * Logica estimate

Figure 5.14
Average Cost per Transaction for a Retail Bank by Delivery Channel.

binding for commercial transactions. The electronic signature bill also enables companies to replace warehouses of paper documents with electronic versions, further reducing their operating costs.

Outlook

For businesses, the increased productivity, improvement in customer service, and reduction in operating costs are truly compelling. This will drive rapid adoption of these services as they become available. The key issues for both businesses and their customers will be speed, reliability, and security. These are technical issues that are well on their way to being solved. Thus, the business market for both voice and data services can be expected to grow dramatically.

VALUE-ADDED SERVICE PROVIDERS . . . .

Value-added service providers seek to use somebody else's network to provide their own service. Examples of value-added service providers include Wireless Internet Service Providers (WISPs), Wireless Application Service Providers (WASPs), telematics (combination of wireless communications and global positioning in an automobile), and telemetry (machine-to-machine or machine-to-human communication).

Wireless Internet Service Providers (WISPs)

WISPs are companies that provide Internet access to subscribers and use somebody else's wireless network to carry their service. This is similar to the wired ISPs like Earthlink and AOL. Most WISPs also supply a unique portal service that enables customers to personalize their home page preferences, like a MyYahoo page, and provides a starting point to access the Internet and a wide range of data.

The complexities of wireless Internet access are numerous, particularly across the patchwork of wireless networks in use in the U.S. This is compounded by the wide variety of devices, operating systems, and interfaces that need to be successfully addressed. Please refer to further discussion of these issues in Chapter 4.

All of the major wireless carriers offer wireless data services and content (also discussed in Chapter 4). However, there is also an emerging market for independent WISPs, or those WISPs that do not own their own network. Independent WISPs are the focus of this chapter.

The major independent WISPs in 2000 were GoAmerica, OmniSky, and Palm.net (which is also discussed in Chapter 3). GoAmerica and OmniSky offer access to the entire Internet, while Palm.net only supports sites that are specifically tailored to its Web clipping service (about 400 sites in 2000).

Additionally, the dominant wireline ISPs/portals, like AOL, Yahoo, and Excite, are seeking to extend their content across the wireless medium. As of early 2001, none of these firms offered wireless ISP access to their customers. However, these firms were partnering with a wide variety of wireless carriers to provision their content wirelessly. As the market for wireless Internet matures, these companies may well enter the WISP market. They will be formidable competitors to the incumbent WISPs, due to their substantial customer relationships, brand equity, and financial resources.

Independent WISPs purchase bulk airtime from providers and then resell it to their customers under a variety of plans. They generally provide access to the entire Internet and reformat the data in a way that takes into account the specific network and device making the information request (except for Palm.net which uses Web clipping for the Palm VII only). This often requires the WISP to strip the graphics and remove repetitive headers and footers from a Web page. This is accomplished through gateway servers.

In 2000, GoAmerica's basic plan was $9.95/month for 25 kb of transmission, and $0.10 per additional kb on the CDPD network, or $0.30 per additional kb on the Mobitex or ARDIS networks. Their unlimited plan was $59.95/month ($49.95/month for the Palm). OmniSky charged $39.95 per month for an unlimited plan and also subsidized the PDA modem to $99 from $299. Palm.net offered a number of service plans, ranging from $9.99 per month for 50 kb to $44.99 for unlimited service (see Table 5.1).

As of the end of the third quarter of 2000, GoAmerica had 30,432 subscribers (Morgan Stanley Dean Witter) versus 25,700 subscribers (Lehman Brothers) for OmniSky, and 121,000 subscribers for Palm.net (Salomon Smith Barney). Morgan Stanley Dean Witter is projecting 1.5 million subscribers for GoAmerica by 2004, while Lehman Brothers is forecasting 1.6 million for OmniSky. Based on current subscribership, bandwidth and handset issues, and expected competitive intensity, these forecasts may prove to be optimistic (Lundberg and Zucker, 2000a, 9; Bensche, 2000b, 3; Gardner 2000, 35).

GoAmerica realized a $38.52 per month ARPU in the third quarter of 2000 versus $31 for OmniSky and $19 for Palm.net. However, both GoAmerica and OmniSky were spending well over $1,000 per gross additional subscriber in 2000. Although customer acquisition cost is expected to moderate over time, it will still remain at high levels, especially in comparison with revenues. ARPU is also projected to decline steadily as competition intensifies. This has led to the expectation among analysts that both GoAmerica and OmniSky will be unprofitable until at least 2003. However, given the aforementioned economic scenario, it is difficult to imagine that these firms will ever generate sufficient profits to earn a satisfactory return on their investments.

In addition to subscriber fees, other potential sources of revenue for WISPs include slotting fees charged to content providers for premium placement of their material, m-commerce commissions, and advertising.

Table 5.1 WISP Comparison.

	GoAmerica	OmniSky	Palm.Net
Subscribers 3Q00	30,432	25,700	121,000
Devices Supported	Palm RIM Pager Pocket PC Laptop PC Wireless Phone	Palm V Handspring Visor	Palm VII
Networks Supported	CDPD ARDIS Mobitex 2G Cellular	CDPD	Mobitex
Unlimited Access Charge	$59.95 (Palm V $49.95)	$39.95	$44.99
ARPU 3Q00	$38.52	$31.00	$19.00

GoAmerica supported many devices and a variety of networks in 2000. The devices supported by GoAmerica included Palm, RIM pagers, laptop computers, and PocketPCs. The networks GoAmerica offered service across included CDPD, ARDIS, Mobitex, and cellular/PCS. OmniSky supported the Palm V and the Handspring Visor in 2000, with plans to extend their service to the HP Jornada in the first quarter of 2001. OmniSky's service was only available on the CDPD network in 2000. In 2000, Palm.net supported the Palm VII over BellSouth Wireless Data's Mobitex network.

Structural Market Assessment

The market structure for independent WISPs can be classified as very challenging (see Figure 5.15). This is due to strong threats from every quadrant of the structural market analysis.

Customer Power

Customer power in the WISP market can be expected to be extremely high as it is in the wired Internet world. In the wired world the price of most content and some access has been driven to zero. Customer acquisition costs are extremely high ($1,000+ in 2000) and the ability to retain customers is unproven.

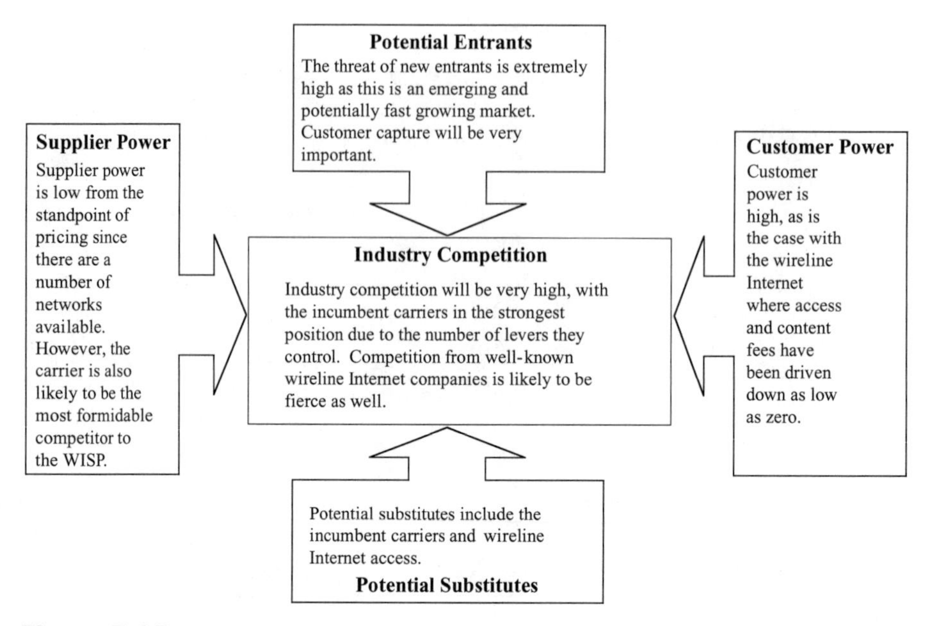

Figure 5.15
Structural Market Assessment: WISPs.

Potential Substitutes

Potential substitutes are an enormous threat to independent WISPs. The main substitutes are the ISP service offered by the carriers and the wired Internet. The incumbents have tremendous embedded advantages over the independents and the demand for mobile Internet service is still unclear, particularly at 2G network speeds.

Supplier Power

Supplier power is likely to be very high since the supplier of connectivity for the WISPs may also be their fiercest competitor. Therefore, prices offered to the WISPs are likely to be higher than the carrier's internal transfer price to its own wireless Internet operations, thus creating a competitive disadvantage for the WISPs.

Undifferentiated content suppliers are unlikely to wield tremendous clout in the industry.

Potential Entrants

Despite the unattractive structure of the market, the threat of new entrants is high as firms from the wired Internet feel compelled to extend their services to the wireless arena. Firms such as AOL and Yahoo will seek to extend their brands and will prove to be formidable competitors if they choose to offer wireless Internet service. Additionally, many small firms will attempt to participate in this market, attracted by the high growth and limited barriers to entry.

Industry Competition

The WISP market is likely to be characterized by intense competition, as is the case in both the wired ISP and wireless service provider markets. This is due to the power of the wireless carriers as both suppliers and substitutes (competitors), the relatively low barriers to entry, the large number of market participants, and the resulting customer power.

Outlook

There is no doubt that there will be a very large market for wireless Internet services; the question that remains is whether the independent WISPs will be able to benefit. The market for WISPs is in its nascent stages and wireless carriers are likely to remain the dominant suppliers of Internet access, due to their inherent advantages. The key service provider advantages, in addition to control of the network and a large installed customer base, include the ability to provide terminals with their own portal as the default, the ability to restrict changing the default portal, the ability to determine which technological upgrades will be made to ensure that their mobile Internet services are optimized for their wireless network, and the ability to influence the speed at which data services are taken up via terminal subsidies.

The carriers were chastened by their experience in the wireline world and are determined not to make the same mistakes in the wireless market. Additionally, once the market has matured more, the firms with large wireline Internet operations can be expected to enter the market aggressively. These companies would include AOL, Yahoo, MSN, and Excite, among others. Since a network is not required to enter this market, this battle is likely to focus on brand, content, and quality of the Internet experience. These are areas where the wireline portals have large advantages over the WISPs.

However, there appears to be room in the market for a well-crafted wireless offering from a noncarrier, particularly if it integrates well with corporate IT systems. In order to prosper, the independent WISPs will have to develop the most compelling wireless experience possible. This will be critical since they are in a much worse competitive position than either the carriers or the wireline Internet powerhouses. Another potential avenue for these firms to pursue, and perhaps the most attractive, is to subcontract their services and expertise to the wired Internet powerhouses or to the carriers, to better enable their offerings.

Wireless Application Service Providers (WASPs)

Wireless Application Service Providers (WASPs) are companies that host or provision applications in a wireless environment. This is a relatively new industry with new firms entering at a rapid pace. As a new industry, the boundaries are not well defined and clear winners have not yet emerged. The services offered by firms in this market space include wireless Web hosting, wireless Web translation, information provision, mobile middleware, and enterprise application hosting. These services are generally managed from a central location or Network Operation Center (NOC).

Market Size and Forecast

This industry is poised to grow rapidly, as the limitations of bandwidth and terminals are successfully addressed and corporations see the key benefits of enabling mobility for both their employees and their customers. However, there are low barriers to entry in this market and competition can be expected to be tough. Additionally, carriers can be expected to enter the particularly lucrative segments of the market, making competition more intense, as suppliers may have to compete with potential customers.

Companies participating in this market generally charge an upfront fee to initiate or configure the service and a monthly fee based on usage or number of subscribers.

A key advantage to using WASPs is that corporate IT departments do not have to devote strained internal resources, or develop costly in-house capability, to gain a wireless presence. There are also lower up-front costs to enabling a wireless platform, and scale economies can be shared with the WASP until traffic warrants an internally managed solution. Some of the scale savings can be servers, gateways, 24/7 maintenance, load balancing, and transport infrastructure.

IDC segments the market into personal (news, weather, stock quotes); collaborative (email, groupware); and enterprise (CRM, ERP, SCM). Enterprise applications, once generally accepted, are forecast to account for the bulk of WASP revenues (see Figure 5.16).

Sampling of Better Known WASPs

Some of the better known WASPs include Aether Systems, AvantGo, 724 Solutions, i3 Mobile, Infospace, and Puma Technology. There are many more companies in this space with more companies entering the market daily.

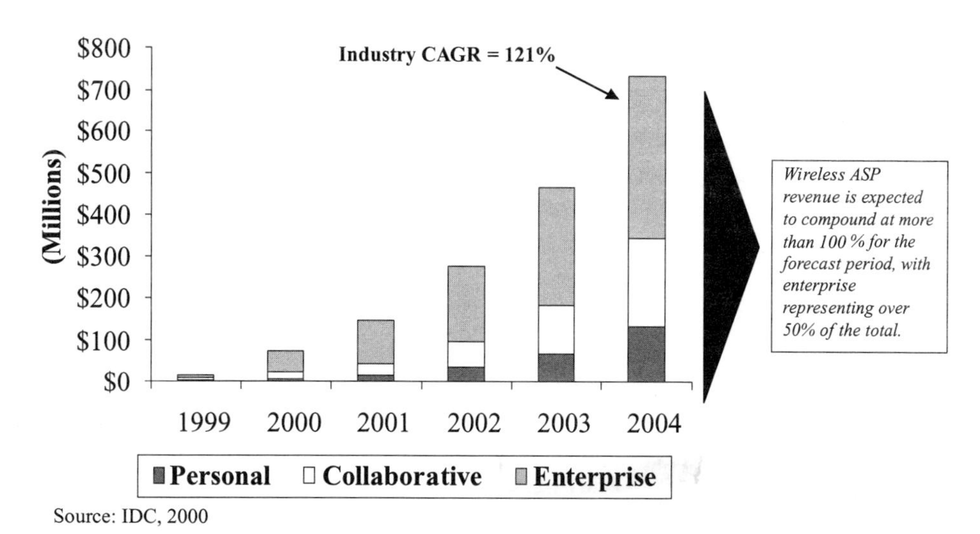

Figure 5.16
U.S. WASP Revenue.

Aether Systems

Aether Systems provides wireless data services, systems, and software that enable people to use handheld devices for mobile data communications and real-time transactions. Aether designs, develops, sells, and supports complete wireless systems for corporations seeking to make data available to mobile workers or consumers.

Aether has focused on developing products for the financial services sector. These services include TradeRunner, a real-time wireless trading and financial information service for Morgan Stanley Dean Witter Online, the Reuters MarketClip service, and several services that deliver financial market information using one- and two-way pagers. Additionally, the company is developing wireless trading and financial services for other major financial institutions, including Charles Schwab.

Aether also owns 33% of OmniSky, a wireless ISP, and acquired Mobeo, which provides major banks and financial institutions with real-time price quotes and news on wireless devices. Aether has made a number of other acquisitions that include LocusOne, which provides wireless data systems to companies that distribute goods and services using their own delivery fleets, and Riverbed Technologies, which develops software that extends the applications and information from corporate networks to handheld devices. Aether has also invested $10 million to acquire a 27.5% interest in Inciscent, a company that plans to develop wireless email, Internet access, and other applications for the small office market.

AvantGo

AvantGo provides software and services that extend the use of Internet-based content and corporate intranets to mobile devices. The company licenses its AvantGo Enterprise products to companies that want to provide their employees, customers, suppliers, and business affiliates access to business information. AvantGo Enterprise Interactive allows two-way communication on mobile devices, including sales force automation, inventory management, and many other applications. AvantGo Enterprise Online allows customers to run these same services without having to operate the AvantGo server software on their company network.

AvantGo's Mobile Internet service allows individuals to access Internet-based content and applications. As of August 31, 2000, AvantGo Mobile Internet service had more than 600 applications and sources of content that were optimized for PDAs and Internet-enabled phones. AvantGo's Mobile Internet service allows the firm to generate slotting fees and advertising revenue and earn commissions on e-commerce transactions.

i3 Mobile

i3 Mobile, Inc., provides timely personalized information to users of wireless communications devices, such as mobile phones, pagers, and PDAs. This service allows users to specify their content and service preferences. i3 delivers data services such as stock quotes, news, weather, sports, and traffic information. Depending on the device and network, a subscriber can send and receive email, update personal calendars, and perform m-commerce transactions.

As of December 31, 1999, i3 had more than 450,000 users of its products and services, with approximately 100,000 paying subscribers and 350,000 complimentary users. The company offers its products and services primarily through cobranded distribution relationships with wireless network operators. Operators offer these products and services to their users on both a complimentary and a subscription basis. These arrangements display both the brand names of the distributor and i3 Mobile.

i3 provides personalized wireless information services to carriers such as AT&T and 15 other wireless network operators. These carriers represented more than 55% of the total North American market of wireless phone users.

i3's interactive wireless portals are designed to provide subscribers with personalized information regardless of the type of wireless device, network infrastructure, or protocol chosen. Information can be accessed either by notification or on request. Messages typically range between 40 characters and 240 characters. Wireless users can interact by two-way Short Message Service (SMS) or through a browser, including a WAP browser.

i3 Mobile provides wireless data services to wireless network operators and corporate enterprise distributors. i3 Mobile acquires content from more than 50 content providers that deliver general, broad-based information, as well as location and user-specific content. i3 is then able to extract the content that is relevant to its users and sends specific messages based on personal profiles.

724 Solutions

724 Solutions provides an Internet infrastructure solution to financial institutions that allows them to offer online banking, brokerage, and e-commerce services across a wide range of wireless networks and devices.

724 Solutions supports digital mobile phones, PDAs, two-way pagers, and PCs through network service providers. The system ascertains the type of device being used and optimizes the information for the particular device being used.

The 724 Solutions Financial Services Platform (FSP) uses an open architecture to deliver secure information, services, transactions, and payments. Their network gateway can translate data into formats appropriate for CDMA, GSM, TDMA, CDPD, and Mobitex networks. This gateway enables services to be deployed quickly and can support new protocols without changing the system architecture.

724 Solutions customers include Bank of Montreal, Bank of America, Citigroup, and Wells Fargo.

InfoSpace

InfoSpace is a leading provider of cross-platform merchant and consumer infrastructure services on wireless, broadband, and narrowband platforms. The company provides commerce, information, and communication infrastructure services to wireless devices, merchants, and Web sites. InfoSpace has relationships with AT&T Wireless, Cingular Wireless, Verizon Wireless, National Discount Brokers, and Bloomberg. InfoSpace's affiliate network also consists of more than 3,200 Web sites, including AOL, Microsoft, GO Network, and Lycos.

InfoSpace has developed a flexible technology platform that enables it to deliver a broad range of services to Web sites, merchants, and wireless carriers. InfoSpace's consumer services are distributed through wireless devices and Web sites. These services are made up of four main components: unified communication services, including email and instant messaging; information services such as sports, news, and information; community services such as online address books and calendars; and collaborative services, such as real-time file sharing.

InfoSpace also offers merchant services for local merchants and distributes these services through regional Bell operating companies (RBOCs), merchant banks, and other financial institutions. These services include commerce services such as building online stores. In 2000, over 350,000 companies used these merchant services.

Through Infospace's acquisition of Millet Software (PrivacyBank.com), Infospace will be able to provide one-click purchases through a wireless device. Infospace's other wireless services include promotions, location-based directory services, and secure wireless commerce through a partnership with VeriSign.

Puma Technology

Puma Technology develops, markets, and supports synchronization, change detection/notification, and Web rendering/browsing software. Through its ProxiNet acquisition, Puma developed Browse-it software, which allows portals, wireless ISPs, and e-commerce companies to provide highly secure, real-time Web access to Palm OS PDAs. Through its NetMind Technologies acquisition, Puma was able to develop Mind-it software through which users can receive instant notification of changes to any content they specify, on either the Internet or corporate intranet. Puma further expanded its capabilities through its Dry Creek Software acquisition and its Mobile Application Platform (MAP) infrastructure for wireless carriers. MAP is the common server platform on which Puma is building its new mobile solutions, and it includes mobile data access engines, a synchronization engine, a personalization and notification engine, and a query engine. MAP is deployed inside the firewalls of major corporations.

Puma also has Intellisync software that allows users to synchronize data on mobile devices with their PCs and groupware servers, through its patented Data Synchronization Extensions Technology (DSX Technology engine). Intellisync automatically monitors all changes in the groupware application through the use of its Notification Transport Processing Technology (NXP Technology engine).

Puma also has a product called Satellite Forms that provides tools to create custom applications for Palm OS PDAs and allows these applications to be integrated with desktop or network databases.

These are just a sampling of the wide variety of firms operating as WASPs. There are many more, and they offer an even wider array of services. There are also more firms entering the market on a daily basis. However, WASPs are generally very small in terms of revenue and are generally unprofitable during their development phase.

Structural Market Assessment

The market structure for independent WASPs can be classified as competitive but constructive (see Figure 5.17).

Customer Power

Customer power is relatively low given the high level of expertise needed to deploy a secure and reliable system. Additionally, internal IT resources are already strained at most potential customers and current traffic does not justify an internal solution. As more firms enter the market and firms gain greater in-house expertise, customer power can be expected to increase.

Potential Substitutes

The largest potential substitute is traditional wired services. These services are well known, reliable, and cost effective in most cases. The key challenge will be to make the business case to extend these services into the wireless arena.

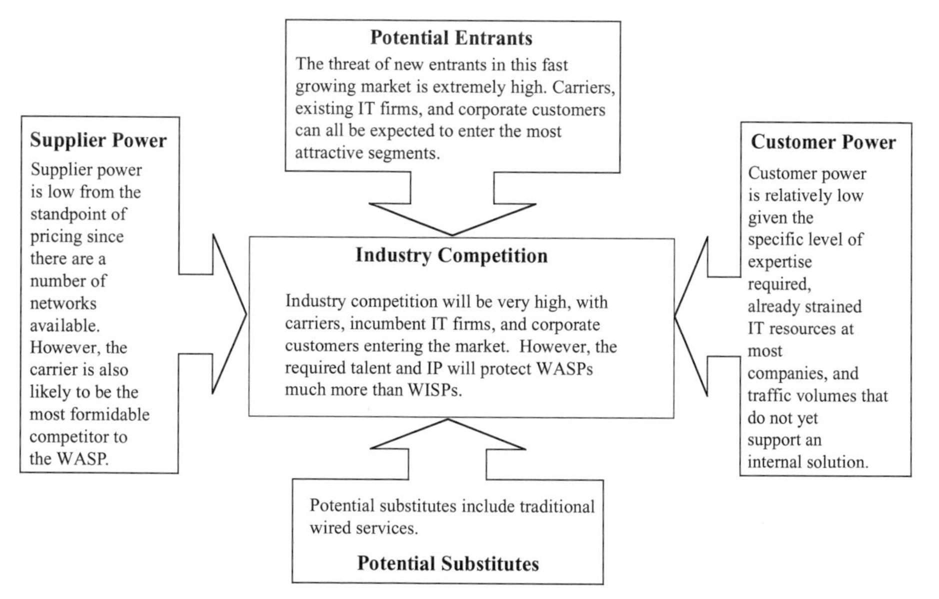

Figure 5.17
Structural Market Assessment: WASPs.

Supplier Power

Supplier power is not particularly high since there are a number of different networks to select from (carriers are more likely to provide more favorable tariffs to WASPs than WISPs and/or the carrier may be the connectivity partner for the WASP). However, the carrier may also prove to be a formidable competitor in the most lucrative segments of the market as they attempt to diversify their revenue streams and avoid becoming "dumb pipes."

Potential Entrants

The threat of potential entrants is very high as new firms enter the market on a regular basis, as corporate IT departments gain expertise and traffic, and as carriers enter the market. The growth rate of the market and the relatively low barriers to entry will attract many firms; however, providing a secure and reliable system is likely to be more complicated than many believe, and experience curves will be very beneficial to early pioneers.

Industry Competition

Competition in the WASP market should be expected to be spirited but probably not destructive. This is due to the rapidly growing nature of the market, the relatively small size of the competitors, and the high level of expertise needed to provision enterprise class services. As more firms enter the market, particularly the larger well-established IT firms and the wireless carriers, the market will become more competitive. However, by that time the superior WASPs will have been able to establish individual reputations.

Outlook

Clearly the WASP market is very exciting, and new capabilities are being enabled on a daily basis. This market is poised for substantial and rapid growth. However, the outlook for WASPs is clouded. The market is in an embryonic stage and many new competitors are entering the market. The current crop of WASPs are small companies with limited capital. They are in their development phase and are consuming capital at a rapid rate, and thus are vulnerable. Many of the companies, mentioned earlier in this chapter, are leading the way in the WASP market, but many firms in this market will not survive, either through acquisition or lack of access to capital.

Carriers, while potentially the largest suppliers to WASPs, may also turn out to be their largest competitors. Additionally, corporate IT departments may use WASP services only until an internal effort is justified and adequate skills are acquired. However, the skills and intellectual property developed by the best WASPs will offer them protection against larger competitors, as this rapidly growing market will be able to support them although only a few will become big winners. The firm or firms that achieve dominance in this space may well become the next Microsoft.

TELEMATICS

Telematics is the integration of mobile wireless communications and global positioning satellite systems (GPS) technologies in an automobile. Telematics enables customers to identify their exact location, monitor the status of their vehicles, receive a variety of information and services, and enable theft-protection measures. These services can be used by businesses to locate and manage fleets of vehicles and by individuals for security purposes, emergencies, and access to information available through the call center. All of the major automakers have begun or implemented telematics programs as of 2001 (see Table 5.2).

Market Size and Forecast

Telematics represents a large opportunity for automobile manufacturers, mobile wireless carriers, and equipment makers. The Strategis Group predicts that telematics subscribers will grow from 4 million in 2002 to more than 17 million in 2005. They also expect telematics to be an option on 84% of new vehicles by 2005 versus just 17% in 2000. They forecast the market for service and equipment to grow to more than $5 billion by 2005. Forrester Research projects 10.4 million telematics systems to be installed in new vehicles by 2003, growing to 18.8 million vehicles in 2004 (see Figure 5.18).

Automakers view telematics as a tremendous opportunity to generate an ongoing revenue stream from their customers and as a means of substantially improving customer care. Roughly 16 million automobiles are sold in the U.S. every year, and there are about 100 million vehicles in service, making telematics a potentially large and lucrative market.

The first telematics program was launched by General Motors, with its OnStar service. This service, which uses a wireless modem and a GPS receiver, provides vehicle monitoring (including a lock-out service, where the doors can be unlocked from the service center), emergency notification when an airbag is deployed, and a variety of concierge services through its monitoring station.

Table 5.2 Telematics Products Offered by OEMs.

Company and Product	Description	Equipment Price	Service Cost
GM OnStar: 1st Generation	First introduced in 1997, phased out in 1999. Includes full-featured cell phone; emergency service; roadside assistance; navigation and vehicle tracking; remote diagnostics and door unlocking.	$895 plus installation (approx. $400).	$22.50 per month, not including cellular service.
GM OnStar: 2nd Generation	3-button system that communicates only with OnStar service centers. Emergency service; roadside assistance; navigation and vehicle tracking; remote diagnostics and door unlocking.	Varies; available only as component in options packages. Was $695 as stand-alone option.	Annual fees of $199 for basic service and $399 for advanced service, including cellular service.
Ford RESCU	Emergency and convenience service available in Lincoln Continentals since 1996.	Approx. $2,000.	No monthly service fee (excluding cellular service charge).
BMW	BMW Mayday Cellular Phone will include 24-hour roadside assistance, route guidance, airbag deployment sensors and remote door unlocking.	$1,200.	Includes 6 months of cellular service, but cellular calling plan is required.
Jaguar	Jaguar Assist will be an option on all model year 2000 sedans; includes on-board navigation system with monitor and full-featured cellular telephone.	$4,300.	Covers service bureau fees, but not cellular service.

Table 5.2 Telematics Products Offered by OEMs (continued).

Company and Product	Description	Equipment Price	Service Cost
Infiniti	Infiniti Communicator is available on the Q45 and I30 luxury sedans. Service includes roadside assistance and navigation. Car doors may be unlocked or locked as needed from a remote location.	$1,600	Includes 4 years of service, but not cellular service.
Mercedes	Telematics unit (Tele Aid) standard on all MY 2000 S-Class, C-Class, CLK, E-Class, CL Coupe and SL Roadster cars.	Standard feature, not priced separately	No service charge; does not include cellular calling plan.

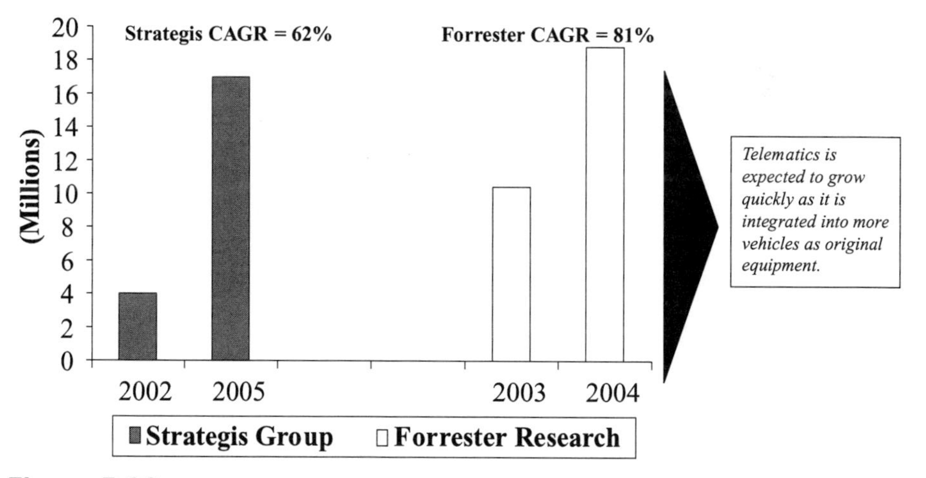

Figure 5.18
U.S. Telematics Subscriber Forecasts.

The OnStar service was introduced as an option on three 1997 model year Cadillacs. By year-end 2000, OnStar was available on 32 of GM's 54 models and had grown to 800,000 subscribers. OnStar will also be available on other manufacturers' products, including Lexus (Toyota) in 2001 and Acura (Honda) in 2002.

GM has added a wireless telephony service and used the power of the vehicle to increase the power of the phone. GM claims that its embedded phone is five times more powerful than the best handsets, transmitting at three watts of power versus 0.6 watts in the most powerful handsets. GM's antenna also has eight times the gain of a normal antenna used inside of a vehicle. This significantly increases the range of the OnStar system versus normal handsets. Verizon is the carrier for this service.

GM has also launched its Personal Advisor service. The Personal Advisor service uses a voice-driven interface to provide personalized Internet data to subscribers. A subscriber can set preferences from a PC and receive the information in the car, audibly, through WAV files. GM will launch a personalized, real-time traffic update service, through Personal Advisor, in the first quarter of 2001.

The fit between mobile wireless communications and the automobile is clear, and research has shown that most wireless phone calls are made from a vehicle (see Figure 5.19). GM claims that drivers in the U.S. spend 500 million hours per week in their automobiles and that 85% of mobile wireless telephony subscribers use their phones while driving (see Figure 5.19). Additionally, an automobile provides the opportunity for a larger display, a constant power source, a keyboard, and increased processing power and storage.

The requirement for a separate phone number and subscription charge for one's car is seen as a serious drawback for telematics. It is unlikely that a large number of customers will be willing to maintain a subscription for both a mobile phone and a telematics program. Ford's Remote Emergency Cellular Unit (RESCU) product tries to bridge this gap by using a removable wireless handset from Sprint PCS as the connection medium.

The desire to maintain only a single wireless communications subscription may lead to a preference for aftermarket systems that use an existing wireless handset.

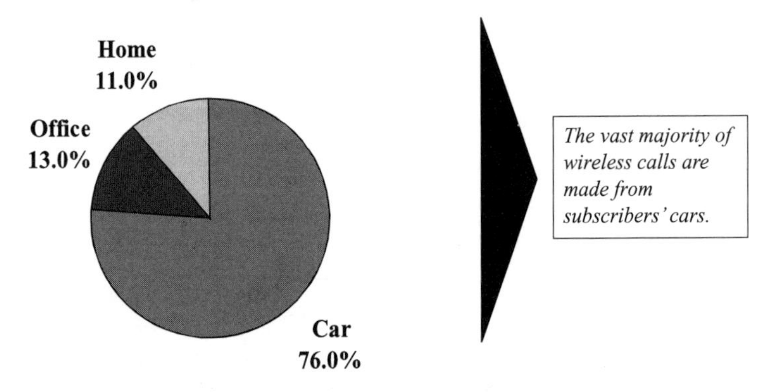

Source: Strategy Analytics, Inc.: a Boston-based company (2000c)

Figure 5.19
U.S. Wireless Subscribers' Calling Location.

However, aftermarket systems are not as well integrated with the major vehicle systems as a factory-installed system.

Additionally, wireless phones or PDAs—in conjunction with voice portals—may also be able to provide some of the same services as telematics. The drawback to this approach is also the lack of integration with major vehicle and safety systems.

Besides safety, one of the key applications for telematics is predicted to be route planning and traffic jam avoidance. Additionally, entertainment—particularly rear seat—is expected to be an important capability. Entertainment will be further enabled once digital satellite radio and digital terrestrial broadcasting are launched in the U.S. This will potentially turn the vehicle into a mobile entertainment platform.

One of the key enablers for telematics will be the development of a robust voice interface. When driving, it is important to minimize distractions. This will require the system to have voice capability, so that the drivers can keep their eyes on the road instead of on a display. This is an important issue that—if not successfully addressed by telematics systems suppliers—will be addressed by lawmakers and trial attorneys. GM's OnStar system deals with this issue by employing a call center with live operators and by using WAV files for their Personal Advisor service, supporting their mantra of "Eyes on the road, hands on the wheel." However, maintaining an operator pool is a more costly solution than using a robust automated voice interface. Unfortunately, the shortcomings of voice-recognition systems, particularly in a noisy automotive environment, have yet to be addressed successfully.

Structural Market Assessment

The market structure for telematics is reasonably favorable, with competition having as much to do with the vehicle as with the telematics system itself (see Figure 5.20). The key structural issue that needs to be addressed is the requirement, in 2000, for a second wireless subscription for the car.

Customer Power

Customer power is moderate due in large part to the high value consumers place on safety. Safety is an important motivator in customers wanting telematics.

If a customer wants to purchase an integrated telematics system with a new vehicle, the choices are limited usually to the manufacturer's own system. Additionally, the manufacturers are beginning to bundle the cost of the system in with the vehicle, so the customer is not aware of the specific cost of the telematics system. However, the cost of a second wireless subscription for the vehicle has dampened consumer demand for telematics. This is one of the most important issues facing the industry.

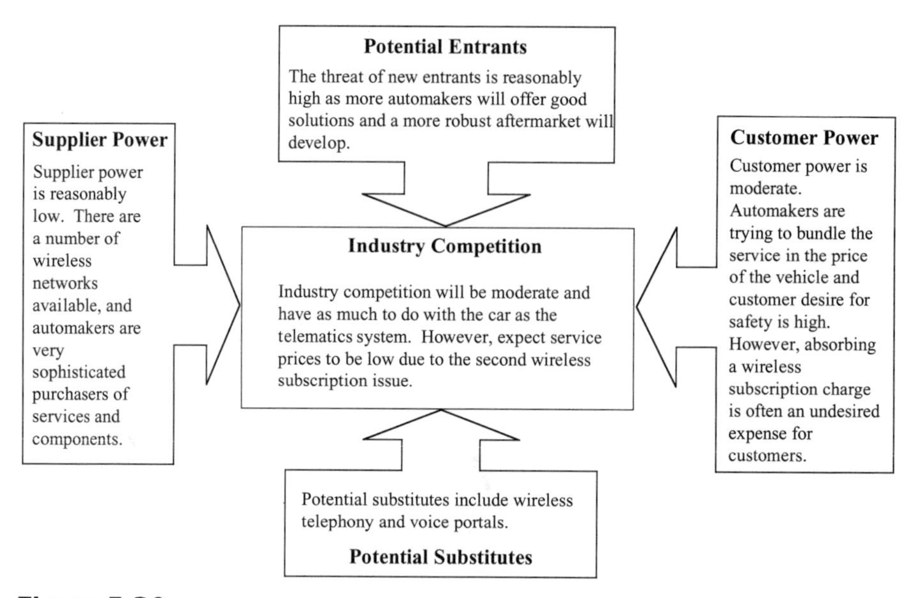

Figure 5.20
Structural Market Assessment: Telematics.

Potential Substitutes

The major substitute for telematics is mobile wireless telephony. This poses a major issue for telematics since the handset is less expensive and is usable outside of the vehicle. However, the handset is not integrated into the major systems of the vehicle (doesn't call for help when airbag deploys) and is not as powerful as the telematics system.

Supplier Power

Supplier power is relatively low since automakers are very skilled at purchasing services and components from suppliers. There are also a large number of carriers to obtain connectivity services from.

Potential Entrants

The threat of new entrants is moderate as more automakers focus on this market and as more aftermarket systems are developed.

Industry Competition

The competition in telematics is inextricably tied to competition in the auto industry. The automotive industry is extremely competitive on a global basis, due to its cyclical nature, high fixed costs (break-even points), and desirability to national governments. We can expect this level of competition to carry over into telematics, especially if this feature is shown to drive sales of vehicles. Additionally, aftermarket systems will be a feature of the market as they are in car audio, though integrated systems are likely to be the preferred solution if they are reasonably priced. However, since telematics is a feature of a larger purchase the competitive dynamics are likely to be muted as compared to the rest of the wireless equipment industry.

Outlook

Telematics is an emerging market with multi-billion dollar potential over the medium term. As bandwidth and applications increase, and the advantages of being integrated with the vehicle (power supply, screen size, power) are recognized, telematics can be expected to grow rapidly. One of the key hurdles to mass adoption of telematics is the subscription charge. If this issue can be addressed through a combination of bundling with the price of the car service provider subsidies, advertisements, and high value-added services, telematics will prosper.

Despite the potential for a substantial aftermarket, similar to the car audio market, the automakers will likely be the main protagonists in the telematics arena. They will drive adoption by bundling the equipment and service into the price of the car. They can also offer a fully integrated solution that is tied into the major systems of the car for monitoring, safety, audio, and appearance.

Automakers will try to use telematics as a way to develop an additional revenue stream and to improve customer care. Entertainment, route planning, and safety are likely to be the main drivers for adoption of telematics, allowing the automobile to become a mobile information platform.

The most interesting feature of the telematics market is that competition will often have as much to do with the vehicle as with the telematics system.

TELEMETRY

Telemetry, for the purpose of this text, refers to machine-to-machine or machine-to-human communications across a wireless network. The main applications of telemetry are monitoring and controlling. Telemetry can be used to monitor the status of vending machines and office equipment or to read utility meters remotely. Telemetry can also be used to control household appliances, security systems, or pipeline switches. Another promising application for telemetry is credit card authorization. Not only can this be done in a mobile environment, but it is faster than a wired connection since a dialup connection does not have to be established in a packet-switched environment.

Telemetry is particularly attractive to businesses whose assets are spread over wide areas, such as vending machines. Coca-Cola has been testing telemetry in vending machines to request refills, alert service personnel to equipment malfunctions, and even to raise prices on especially hot days. This can lead to substantial increases in the financial performance of the vending machine, as out-of-stock situations are avoided and revenues are maximized during peak demand periods.

There are two principal monitoring methods for telemetry—constant communication and periodic communication. Constant communication provides a continuous stream of data, and periodic communication transmits at predetermined times or when certain conditions are met.

Telemetry can communicate over cellular networks, paging networks, and data networks. However, due to superior coverage, in-building penetration, and cost, paging networks may be the most promising. Additionally, paging networks tend to have excess capacity, which is not always the case with cellular networks. To this end, Motorola has developed the CreataLink 2 XT data transceiver that operates on their ReFlex network.

Market Size and Forecast

In 2000, telemetry was a relatively small market due in large part to pricing of the hardware. However, the market may be expected to grow as the cost of transceivers and telemetry modules falls. The key to success in telemetry will be in deploying many units that require frequent or constant communication.

According to Strategy Analytics, telemetry is poised to grow rapidly off of a small base (see Figures 5.21 and 5.22). It is important to note that their forecast considers only cellular telemetry applications and does not include paging and data networks. This indicates that there is likely to be a larger overall opportunity than is depicted in these charts. Additionally, as paging network operators continue to face

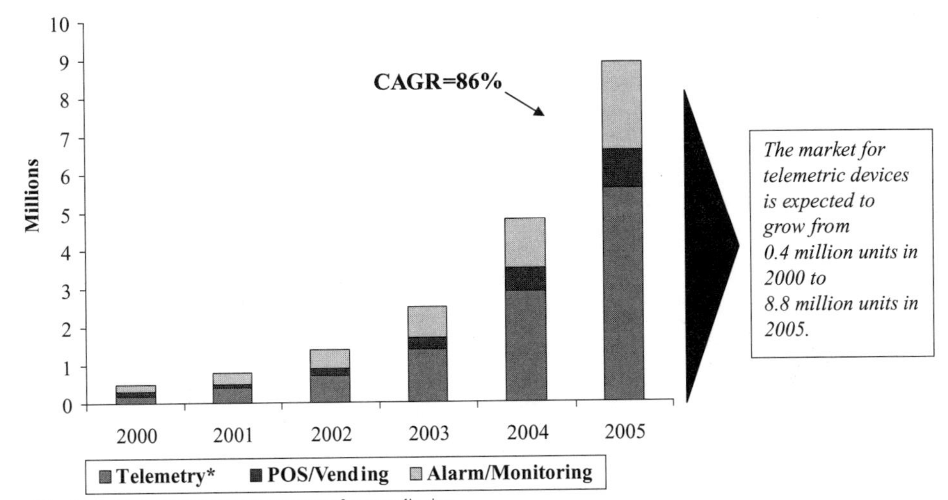

* For forecast purposes, the telemetry category refers to applications such as utility meter reading and other residential remote controlling

Source: Strategy Analytics, Inc.: a Boston-based company (2000e)

Figure 5.21
U.S. Telemetry End Points.

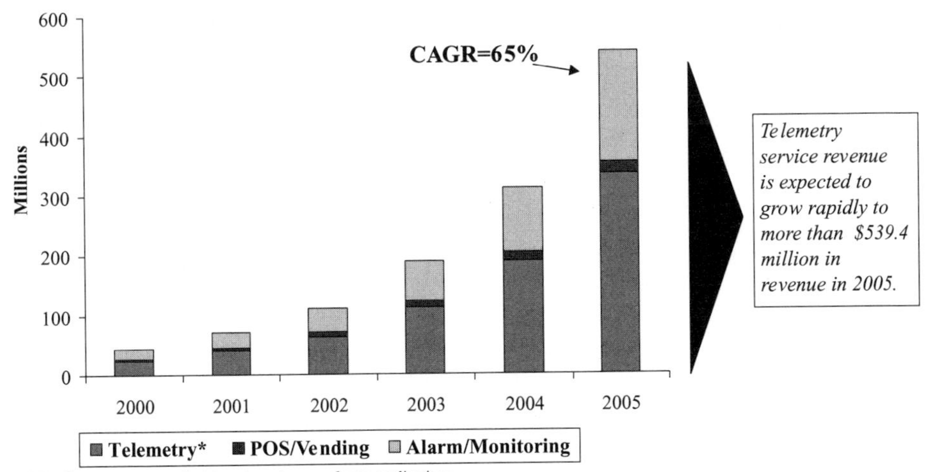

* For forecast purposes, the telemetry category refers to applications such as utility meter reading and other residential remote controlling

Source: Strategy Analytics, Inc.: a Boston-based company (2000e)

Figure 5.22
U.S. Telemetry Service Revenue.

difficult market conditions, they can be expected to aggressively try to expand the business opportunities for their networks. Given the inherent advantages of paging networks, they can be expected to meet with some success. This is especially true for in-building machines like office equipment and vending machines.

Outlook

Although there are only modest expectations for telemetry, the industry may well prove to be more attractive than is being forecast. If a reasonably priced mass-market solution can be deployed, the large benefits in terms of service staff efficiency and customer satisfaction are likely to be too compelling for businesses to ignore. Many customers would be thrilled to see an equipment repair person appear when a piece of equipment is near failure, rather than having to call in a repair person once it has failed. Additionally, dispatching a vending service person with the correct items at the appropriate time will allow for more service calls and better route planning. The cost of meter reading will also be substantially reduced, and utilities can implement more sophisticated load planning techniques if they have real-time consumption data. In the future, many more useful telemetry applications will be developed.

In terms of cost, the effective use of telemetry can automate many very labor-intensive tasks, leading to substantial reductions in service staffs. Since personnel costs are one of the largest cost components of these businesses, increased efficiency can lead to a markedly lowered cost structure.

These benefits are likely to prove persuasive to businesses and propel telemetry to above forecast growth rates.

Epilogue

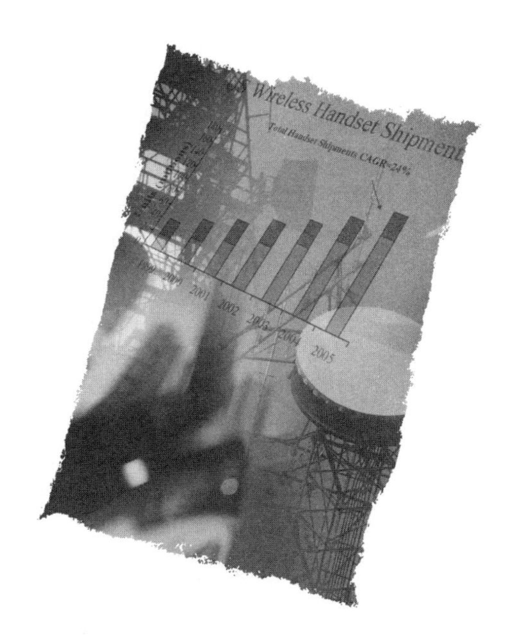

As we go to press, roughly a year after becoming interested in the mobile wireless communications industry, much has happened. Most notably the NASDAQ has plunged from above 5,000 to below 2,000. This descent has impacted telecommunications companies more than most, and, based on their inflated valuations versus their legitimate business prospects, should have been expected. The hysteria surrounding the European 3G auctions in the summer of 2000 marks the high-water mark for the industry and the valuation of the NASDAQ. Analysts tried to figure how the auction prices could possibly be justified, but couldn't, and this affected a wide swath of the technology sector as profit expectations had to be scaled back sharply. The debt used to secure these licenses now calls into question the ultimate viability of some of these carriers and, consequently, capital in the industry has been severely rationed, particularly to the less capable firms.

However, as noted in the introduction, the mobile wireless communications industry is ultimately propelled by the three primary growth drivers of the U.S. economy: digitalization, the speed and power of information processing, and mobility. This foundation, along with an improving political climate and the general economy's current financial difficulties (which should weed out the less capable companies), is actually serving to improve the long term structural attractiveness of the industry in the U.S.

And again here in May 2002 →

The 700-MHz auction has been postponed again, and the projected proceeds have been taken out of the budget submitted by President Bush, for the next several years. This restricts the supply of spectrum. By maintaining the status quo, confusion in the marketplace will be reduced, as will the capital requirements for carriers and R&D expenses for the handset and infrastructure suppliers.

The FCC is seriously considering altering the spectrum caps to increase the amount of spectrum that one company can control in a market. If this should come to pass, it would become the single most important factor in improving the attractiveness of the industry, as the industry would consolidate around the strongest players and pricing would tend to stabilize.

As expected, the handset market is beginning to consolidate as the financial and competitive realities become more apparent. Ericsson has withdrawn from the manufacture of handsets through a relationship with Flextronics and an alliance with Sony. Further action can be expected from Philips and possibly Siemens as their businesses continue to struggle. Handsets offering higher data rate services have been far more difficult to develop than initially expected which is likely to make it more difficult for second tier firms to keep pace with the leaders. Still, handset unit sales growth in the first quarter of 2001 was in excess of 10%—much faster than the growth of the overall economy which highlights the systemic growth characteristics of the industry.

The infrastructure industry has suffered as well, as capital constraints have reduced the rate of growth in network construction. In Europe serious consideration is being given to network sharing, in which one infrastructure would be built with several firms sharing it until demand is more certain (this is unlikely to be a viable long-term solution). Additionally, the engineering tasks have proved more complicated than the industry expected. For example, in Japan where there is very tight control over standards, DoCoMo has had to push back the launch of their W-CDMA network for four months until the fall of 2001 to deal with network issues like call handoffs. Unfortunately, history has shown that when projects of this nature begin to slip, they generally continue to slip. This will place a greater premium on 2.5 G and 3G infrastructure suppliers who develop exceptional systems integration and network planning capabilities.

Despite the tremendous short-term pain for the industry, a more structurally sound industry structure appears to be emerging in the U.S., where there are fewer but stronger carriers with reduced capital obligations, fewer handset suppliers dealing with more complicated products to manufacture, and more capable infrastructure suppliers with large, complicated integration and network tuning projects. Additionally, company share prices have been marked down substantially, which may lead to further industry consolidation at more reasonable prices. The growth of the industry will continue to be fueled by the fundamental drivers of the economy and the compelling advantages of the service, although prices may not decline as quickly as they have in the past. This points to a more positive industry environment in the U.S. as we migrate to 2.5G and 3G services.

Glossary

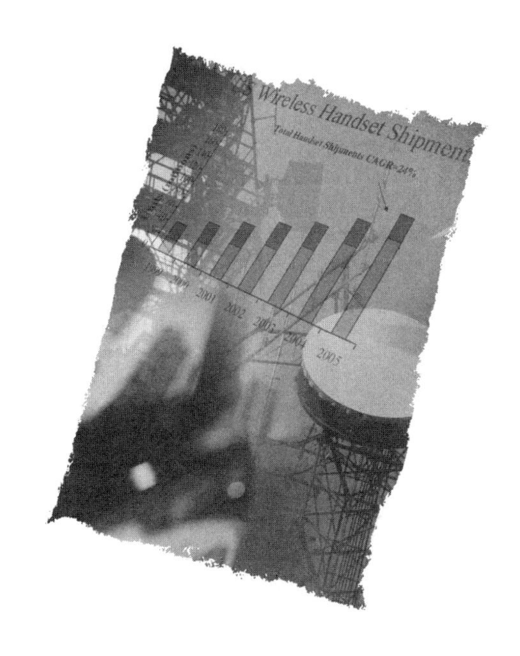

1G

First-generation or analog transmission technologies. Analog technologies use spectrum or bandwidth very inefficiently.

1XEV (also known as HDR [High Data Rate])

Intermediate CDMA solution from Qualcomm; estimated to support transmission speeds in excess of 2 Mbps.

1XRTT

Part of the CDMA 2000 standard and a relatively straightforward upgrade to CDMA IS-95, requiring a software and chip board upgrade to the network. According to Yankee Group, this is likely to cost about 30% of the original network cost, which is substantially less than the cost of upgrading other networks and has the advantage of substantially increasing voice capacity. The 1X refers to 1.25 MHz of operating bandwidth and RTT is short for Radio Transmission Technology. 1XRTT is expected to provide a packet data rate of 144 kbps in a mobile environment.

1Xtreme

Proposal from Motorola and Nokia that is reputed to support peak speeds over 5 Mbps and steady state speeds of 1–2 Mbps.

2G

Second-generation, or 2G technologies, are digital circuit-switched systems, requiring dialing and establishing a dedicated circuit between the caller and the receiver. This dedicated circuit is maintained for the duration of the call, regardless of the actual quantity of information transmitted. This is similar to the current landline or POTS (Plain Old Telephone System) phone system.

3G

Third-generation, or 3G technologies, are broadband packet-switched technologies. Generally, 3G refers to packet-switched transmission rates up to 144 kbps in a mobile vehicular environment, 384 kbps in a mobile pedestrian environment, and up to 2 Mbps in a stationary environment. 3G should also provide multimedia capabilities, location information, and wireline voice quality and security, and support simultaneous voice and data connections.

3XRTT

Second phase of the CDMA 2000 standard, promises data rate increases to 2 Mbps, and will use 3.75 MHz (3X) of bandwidth. RTT is short for Radio Transmission Technology. Commercial availability for 3XRTT is expected in 2002. With the emergence of higher speed intermediate solutions, however, the need to move to 3XRTT may be reduced.

Angle of Arrival (AOA)

Uses an array of large antennas at the cell site to measure the angle of the incoming control signal from the handset. A minimum of two cell sites is required to determine location, and no handset modifications are needed. However, this solution requires substantial capital expenditures by the carrier, particularly in rural areas. AOA accuracy is negatively impacted by line-of-sight obstacles and distance from the base station and may not function well in an urban environment.

Assisted GPS (A-GPS)

By shifting much of the processing burden from the handset to the network, A-GPS helps to overcome some of the drawbacks of pure GPS such as cost, power consumption, speed to determine location, and the line-of-sight requirement. Additionally, the network keeps track of location so that when satellites are obstructed, a good estimate of location can be obtained based on the last reading. A-GPS is not costly (~$20) from a handset perspective but requires additional investment in the network. Location can usually be ascertained in about 5 seconds and A-GPS accuracy is considered the highest.

Authentication Center (AuC)

Database that keeps the authentication register of all subscribers. The mobile handset contains a key that must be authorized by the AuC for the handset to gain access to the network.

Average Revenue per User (ARPU)

Refers to the average monthly bill per subscriber.

Bandwidth

Breadth of the frequency used. Analogous to a water pipe, in which a larger diameter allows more water to be moved. Therefore, a 30-MHz wireless license is much more valuable, in terms of capacity and throughput, than a 15- or 10-MHz license.

Base Station Controller (BSC)

Hub of the wireless infrastructure. It handles all connections with moving mobile terminals. The BSC provides the connection between the mobile system and the MSC. The BSC is responsible for channel setup, frequency hopping, and call handoffs (switching channels in the same cell and switching cells under the domain of that particular BSC). The base station controller can connect as many as several hundred base station transceivers.

Base Station System

Provides the interface between the mobile handset and the wireless network. This system is the most costly part of the wireless infrastructure and can represent about 70% of infrastructure hardware cost.

Basic Trading Areas (BTAs)

Regions that include a single metropolitan area. The PCS spectrum was divided into six segments to be auctioned: the A, B, C, D, E, and F blocks. The C (30-MHz) and D through F (10-MHz) blocks provided coverage for the 493 BTAs.

Bluetooth

Royalty-free global wireless technology standard. In early 1998 a group of computer and telecommunications companies (including Intel, IBM, Toshiba, Ericsson, and Nokia) began to develop a way for users to connect a wide range of mobile devices without cables. This standard quickly gained new members, including 3COM/Palm, Compaq, Dell, Lucent, and Motorola. Bluetooth membership numbered over 2000 by the end of 2000.

Bluetooth uses tiny radio transceivers that operate in the unlicensed 2.4-GHz band to transmit voice and data at up to 721 kbps. Bluetooth communicates within a 10-meter perimeter and does not require line of sight to establish a connection. The cost of the transceiver is about $20 and should eventually fall to $5.

Business consumers

Highly valued by carriers since they spend more on mobile wireless services and churn less frequently than retail consumers.

Calling party pays

Practice in which the calling party pays for the call to a wireless subscriber. In Europe a wireless phone number is identified by a prefix that alerts the caller that they are calling a wireless phone and that the call is subject to a charge. In the U.S. the wireless subscriber usually pays for both calls placed and calls received by a wireless phone. U.S. carriers blame the lack of calling party pays for reducing utilization of wireless communications and hindering the adoption of wireless phones as a replacement for landline telephones. The evidence has shown that customers are hesitant to give their phone number out and often do not turn their handsets on in order to reduce charges. This has suppressed minutes of use and monthly charges by up to 20%. This may become less of an issue as low-cost, large bucket-of-minute rate plans become the norm.

CDMA2000

An evolution to 3G technology for CDMA networks. CDMA2000 is compatible with current CDMA networks, IS-95A and IS-95B (data-enabled CDMA). CDMA2000 is expected to be rolled out in two phases: 1X or 1XRTT and 3X or 3XRTT. It is also forecast to lead to a twofold increase in voice capacity and battery standby time. Commercial availability is expected in 2001.

Cell of Origin (COO)

System currently used to comply with Phase I E-911 requirements. This technology tells which cell a caller is occupying, but offers no greater resolution. Cell location can generally be ascertained in about 3 seconds. The accuracy of COO is determined by the teledensity of the area, with accuracy proportional to the number of cell sites, or size of the cell. This solution requires no alteration to the network or to the handsets, but is insufficient for emergency services.

Cellular Digital Packet Data (CDPD)

Packet-switched data transmission technology developed for use on cellular frequencies. It is considered a precursor to GPRS. CDPD is a cellular network overlay that utilizes unused channels (in the 800-MHz to 900-MHz range) to transmit data in packets. CDPD offers data transfer rates up to 19.2 kbps.

Cellular service

Refers to mobile wireless telephony at the 824–849 MHz and 869–894 MHz frequency bands.

Cellular structure

Describes the wireless telephony network as a patchwork of hexagonal cells, like a honeycomb. Each cell contains its own radio equipment and is allocated its own set of voice channels. Adjacent cells are assigned different channels in order to avoid interference. Nonadjacent cells, however, are able to reuse the original channels, thus increasing capacity. As the network gains more subscribers, the power of each transmitter can be reduced so that cells can be located closer together, reusing the spectrum more frequently.

Churn

Describes wireless service disconnection. This can happen at the request of the customer or in the event of nonpayment for services, at the request of the carrier. Churn figures are generally quoted on a monthly basis. There are two types of churn: in-system, where someone disconnects and reconnects service with another carrier, and out-system, where someone disconnects service and doesn't reconnect. Out-system churn includes people who voluntarily cease service, move out of the area (and may reconnect at new location), or are disconnected by the carrier.

Code Division Multiple Access (CDMA)

Spread-spectrum approach to digital transmission. This method of transmission assigns unique codes to each transmission and then transmits over the entire spectrum. The mobile phone is instructed to decipher only a particular code in order to receive the designated transmission. The assignment of separate codes allows multiple users to share the same air space. CDMA increases capacity 8–20 times vs. analog cellular and also has greater capacity than TDMA.

Communication tower

Holds the carrier's radio antenna, or base station. Towers are very capital and maintenance intensive.

Customer acquisition cost

Cost for a service provider to acquire a single customer. The major components of acquisition cost are equipment subsidies, sales commissions (at company store or reseller), and marketing costs.

DataTac

Developed by Motorola and IBM to give IBM service employees remote access to the company mainframe. The service, named ARDIS, commenced in 1990. ARDIS is similar to Mobitex in that it is a digital two-way packet-switched data network. ARDIS supports data speeds up to 19.2 kbps and covers 80% of the U.S. population, including Puerto Rico and the Virgin Islands.

Digital transmission technologies

The method through which radio signals are transmitted and received between the handset (terminal) and the base station. The use of different digital transmission technologies is one of the key differences between the U.S. and European markets. The three major digital transmission technologies employed in the U.S. are CDMA, TDMA, and GSM. IDEN, a proprietary technology from Motorola that is similar to TDMA and GSM, is used by Nextel.

Enhanced Cell ID (E-CID)

A software-based solution that determines location by comparing the list, or table, of cell sites available to the handset. Once the available cell sites are known (this is constantly updated), location can be calculated based on the intersections of the overlapping cells. This system works best in areas with many cell sites, and location can be determined within about 100 meters (250 meters in rural areas). A key advantage of E-CID is that line of sight is not required. Currently, this system only works with GSM networks. Because E-CID requires only slight modification to the SIM card in the handset and a proprietary network server it is regarded as a relatively low-cost and nondisruptive solution for GSM operators.

Enhanced Data Rates for GSM Evolution (EDGE)

An evolutionary path to 3G services for GSM and TDMA operators. It represents a merger of GSM and TDMA standards and builds on the GPRS air interface and networks. EDGE is a data-only upgrade and supports packet data at speeds up to 384 kbps. EDGE is able to achieve increased data transmission speeds through a change in its modulation scheme, from GMSK to 8 PSK. The upgrade to EDGE is relatively expensive and requires that carriers replace the tranceivers (radio antennas) at every cell site. According to Commonwealth Associates, this can cost as much as 60% of the original network cost.

Enhanced 911 (E-911)

An FCC mandate that wireless carriers provide location information on callers from wireless handsets. E911 is to be implemented in two phases. Since April 1, 1998, Phase I of Enhanced 911, the FCC has required wireless carriers to transmit all wireless 911 calls to emergency assistance providers operating Public Safety Answering Points (PSAPs). The wireless carriers are required to provide the telephone number and cell site information to the PSAP. This allows the PSAP to call the individual back if the call is disconnected and gives a rough indication of the caller's location. Location by cell site is not precise enough to adequately dispatch emergency services, but does provide neighborhood/town scale location information. Phase II of the FCC's E-911 rules requires wireless carriers to have Automatic Location Identification (ALI) capability in order to pinpoint caller location more precisely. Phase II is supposed to be operational by October 1, 2001. The FCC does not require a specific technology to be used to meet its mandate, but does require adherence to performance metrics. Carriers may employ a location technology that is either network- or handset-based as long as the technology employed meets the following standards for accuracy and reliability. For network-based technology, location accuracy must be within 100 meters 67% of the time and within 300 meters 95% of the time. For handset-based technology, location accuracy must be within 50 meters 67% of the time and 150 meters 95% of the time.

Enhanced Observed Time Difference (E-OTD)

Operates under the same principles as TDOA (measuring the time it takes to receive a signal), but in reverse. The signals are received from at least three base stations, whose locations are known, and location is calculated by the handset. E-OTD utilizes the existing capabilities of the GSM protocol and is relatively straightforward to apply to these networks. E-OTD is a more costly and complex solution to deploy TDOA because of handset software upgrades and location measurement units (reference beacons), but it yields much better location information. Location measurement units are distributed throughout the network, with about one unit for every four cell sites. Time keeping is of the utmost importance, and system time is usually kept by an atomic clock. E-OTD can usually provide location information accurate to 50–125 meters within 5 seconds. However, this system can be susceptible to distortion in urban areas.

Enhanced Specialized Mobile Radio (ESMR)

Represents spectrum located in the 800- and 900-MHz bands. ESMR spectrum was allocated by the FCC in a piecemeal fashion. Use of the spectrum was basically offered on a first-come first-served tower-by-tower basis. Subsequently, much of this spectrum was acquired by Nextel.

Equipment Identity Register (EIR)

Contains information about all valid mobile terminals on the network.

FCC

U.S. federal agency responsible for managing interstate and international communications, including telephony, wireless communications, television, radio, etc.

FLEX

A wireless paging technology developed by Motorola and introduced in 1993. FLEX improves channel efficiency, reduces the cost of paging networks, and enables new services. FLEX enables data speeds of up to 6.4 kbps. FLEX and its derivatives command over 90% market share. FLEX and ReFLEX are not packet-based systems.

Frequency

The place on the electromagnetic spectrum where a signal is transmitted or received. At higher frequencies, more information can be transmitted, due to the shorter wavelength. However, the shorter wavelength decreases the travel distance of the signal and increases the likelihood of interference. Higher frequencies also require larger and more expensive components.

General Packet Radio Service (GPRS)

A 2.5G technology that allows networks to send packets of data at rates up to 115 kbps. GPRS allows "always on" connections to send information immediately to the subscriber, with no dial-up required. GPRS is more efficient than sending data over a circuit-switched wireless connection and allows users to be charged per packet of data rather than by connection time. GPRS is a data-only packet network overlay for GSM networks and is a relatively straightforward upgrade to existing networks, requiring a software and chip board upgrade. GPRS also requires a Gateway GPRS Support Node (GGSN), which is a packet router, and a Serving GSN Support Node (SGSN), which tracks the subscriber and provides security.

Global Positioning System (GPS)

A world-wide radio-navigation system comprised of 24 satellites and ground stations sponsored by the U.S. Department of Defense. The system measures the longitude, latitude, and elevation of the receiver. Triangulation is used to determine location. This is accomplished by measuring the time it takes to communicate with 3 satellites. A fourth measurement is taken to ensure that the timing of the pseudorandom codes is synchronized. Because time is critical in the calculation of location, an atomic clock is used. GPS requires line of sight in order to calculate location. This means that location cannot be determined if the user is inside a building, in an urban canyon, or under a heavy canopy of trees. To communicate with the satellites and to perform the complex calculations, GPS takes the most time of the location technologies to determine location, requiring 10–60 seconds.

Global System for Mobile Communications (GSM)

A similar transmission method to TDMA, in that it assigns time slots to communications. GSM is the standard adopted by Europe and much of the rest of the world, but in 2000 was the least widely deployed of the three major technologies in the U.S. GSM is a complete standard, as it governs features in addition to the air interface, such as SMS messaging, and enables certain location technologies such as E-OTD and E-CID.

High-Speed Circuit-Switched Data (HSCSD)

Offers speeds of up to 58 kbps through a software upgrade to GSM networks. HSCSD is not a packet-switched technology, but works by combining transmission channels to improve data transmission. HSCSD's major disadvantage is a reduction in network voice capacity. Its limitations make HSCSD an unlikely evolutionary path, given the more robust capabilities of GPRS. No U.S. carrier has announced that it will be implementing this solution.

Home Location Register (HLR)

Centralized database that stores information on all subscribers. The HLR also maintains location information about the subscriber.

IDEN

Proprietary digital transmission standard developed by Motorola and used by Nextel. This standard is similar to TDMA and GSM because it uses time division to increase spectrum efficiency.

i-Mode

Mobile wireless data service from NTT DoCoMo that was launched in Japan on February 22, 1999, and signed up more than 12 million subscribers in the first 18 months. i-Mode uses the cHTML (Compact HTML) language and a packet-data (9.6 kbps) transmission system, where customers are charged according to the volume of data transmitted, not the time spent on line.

Infrastructure

The equipment that enables wireless communication between mobile terminals and other terminals, either mobile or landline. This infrastructure consists of several key elements: the switching system, the base station system, and the operation and support system.

Interactive paging

Also known as 2-way paging. The classes of interactive service can be defined by their level of interactive capability. These classes include 1.5 way, where the device sends a receipt or confirmation back over network; 1.7 way, where the device has 2-way communication with optional preset responses; and full 2-way pagers that transmit original messages and responses with a QWERTY keyboard.

Location technology

The ability to identify the location of a mobile terminal, and hence the subscriber, with a reasonable degree of accuracy, as it moves throughout the mobile wireless network coverage area.

Major Trading Areas (MTAs)

Regions that include multiple cities or states. The PCS spectrum was divided into six segments to be auctioned: the A, B, C, D, E, and F blocks. The A and B blocks were 30 MHz each in the 51 Major Trading Areas.

m-commerce

Commerce transacted through a mobile Internet connection.

Metropolitan Service Areas (MSAs)

The FCC divided the county into 306 MSAs and 428 Rural Service Areas. Each MSA and RSA was to have two cellular service providers in order to foster competition. One 25-MHz license was given to the local wireline company (B Block license), and the second was assigned to a nontelephone company (A Block license). The local wireline company was generally the Regional Bell Operating Company such as Bell Atlantic or Bell South.

Minutes of Use (MOU)

Minutes of use is a typical measure of wireless telephony usage for carriers.

Mobile Switching Center (MSC)

Provides the traditional switching function of the landline phone network. This function is to open a dedicated circuit between the caller and the receiver. The major suppliers base their mobile switch on their landline switch. Thus, Lucent bases its mobile switch on its 5ESS switch. The MSC provides the connection to the Base Station System and to the landline network and sets up calls to other MSCs. The MSC also handles call handoffs not handled by the BSC (calls under different BSCs but the same MSC and calls under different MSCs). Each MSC can handle about 100,000–150,000 subscribers.

Mobile wireless data/Internet

The transfer of nonvoice data in a mobile wireless environment. This data can include text, images, audio, video, or large files. Mobile wireless data and Internet access represent the next major growth drivers for the wireless industry, and these services are counted on heavily by wireless carriers.

Mobitex

A digital two-way packet-switched network protocol that operates on cellular frequencies to deliver packetized data. This standard was developed by Ericsson and Swedish Telecom Radio in the early 1980s. The Mobitex standard was subsequently opened up to other manufacturers, in 1994, to broaden its appeal. Nortel, Lucent, and Hughes Network Systems have entered the market and supply equipment based on this protocol.

Modem

A device that converts signals from analog to digital (modulation) and digital to analog (demodulation).

Multipath Fingerprinting (MF)

Locates a caller by matching the received radiowave to a reference radiowave in the system database. The reference radiowave, or fingerprint, stored in the database takes into account the wave reflections generated by making a call from a specific location. This fingerprint is then matched to the actual call, and the location is estimated. Due to the dynamic nature of the environment, maintaining an up-to-date database is critical. Only one cell site is needed to determine location and no handset modifications are required. MF is regarded as the most accu-

OFDM - Orthogonal Frequency Division Multiplexing

rate and fastest network-based system; it can usually fix location within 86–100 meters in 1–2 seconds. This system is regarded as fairly inexpensive to deploy.

Network infrastructure

Equipment that enables wireless communication between mobile terminals and other mobile or landline terminals. This infrastructure consists of several key elements: the switching system, the base station system, and the operation and support system.

Network service providers

Also called carriers, these firms have secured spectrum and built a mobile wireless communications network. The largest mobile wireless communications market in terms of revenue is cellular telephony.

Operating system

The software that governs the operation of the device. For PDAs the most popular operating system is the Palm OS, followed by Pocket PC. Wireless handsets also have operating systems.

Operation and support system

Manages the network. In order to optimize network performance, the system handles system and cell planning, generates traffic reports, and measures radio signal.

Packet switching

A packet-switched system transmits data in packets that are reassembled by the receiver, rather than by establishing a dedicated connection. This is the same method used by the Internet. Packet switching allows for an "always on" connection and a vast increase in capacity, since a dedicated circuit contains much more bandwidth than is consumed by a voice call. Packet switching is a critical enabler for most of the data services currently being contemplated, and it allows for the "push" of information to the customer. Packet-switched systems such as GPRS were being piloted in Europe in late 2000.

Pagers

Devices used to receive paging network signals. These signals can be tone, numeric, or alphanumeric. Traditionally, paging has been a one-way medium. However, with Mobitex, ARDIS, and the advent of ReFLEX, interactive mobile data communication has been enabled. This allows customers to send and receive e-mail through their pagers, using a QWERTY keyboard.

Personal Communications Services (PCS)

Refers to mobile wireless telephony at the 1850–1990 MHz spectrum band. The PCS spectrum was divided into six segments to be auctioned: the A, B, C, D, E, and F blocks.

Personal Digital Assistants (PDAs)

Small hand-held devices that include processing and storage capability and generally run on the Palm OS or Windows CE software platforms. The PDA includes a PIM and a variety of other applications, developed both by the manufacturer and by the developer community. Currently, PDAs can have integrated wireless capability, an add-on wireless module, or no wireless connectivity at all.

Power

The energy in the wave. When power is increased, travel distance increases and the likelihood of interference is reduced.

Prepaid plan

A wireless service plan in which the customer pays for service before using it, versus the traditional plan in which service charges are billed. This is an attractive option for customers who have credit problems and for people who want to regulate spending. As many as 30% of applicants for wireless service are turned down, due to impaired credit. This differs from landline telephony which has a mandate to provide universal service, due to its lifeline status.

Public Safety Answering Points (PSAPs)

Facilities operated by emergency assistance providers to which the FCC has required wireless carriers transmit all wireless 911 calls. Funding for the PSAPs is usually collected by the carriers through a 911 surcharge, which varies from state to state, but is usually $0.70–$1.00 per month per customer.

Public Switched Telephone Network (PSTN)

The traditional landline phone system operated by the Bell companies.

ReFLEX

ReFLEX is a follow-on to FLEX that enables two-way alphanumeric paging through a response channel. ReFLEX enables an immediate response from the terminal and guarantees message delivery. Two versions of ReFLEX are available: ReFLEX 25 and ReFLEX 50. ReFLEX 25 receives data at 9.6 kbps and transmits at up to 6.4 kbps. ReFLEX 50 receives data at 25.6 kbps and transmits at up to 9.6 kbps.

Regional Bell Operating Company (RBOC)

Local phone companies, which were formed as a result of the 1982 breakup of AT&T. RBOCs provide local telephone service to the customers in their regions, but they were not allowed to offer long-distance service. The original seven RBOCs were NYNEX, Bell Atlantic, BellSouth, Ameritech, USWEST, Southwestern Bell, and Pacific Telesis. Through mergers, the four remaining RBOCS are SBC (Southwestern Bell, Ameritech, Pacific Telesis), Bell Atlantic (Bell Atlantic, NYNEX), BellSouth, and Qwest (Qwest, USWEST). In 1996, the Telecommunications Act allowed the RBOCs to enter the long-distance market, provided they could demonstrate local competition. In 2000, the vast majority of areas in the U.S. were not offered long distance by their RBOC.

Retail consumers

Individual subscribers who use wireless services primarily for communication, convenience, and safety. Retail consumers comprise 65–75% of subscribers to mobile wireless communications.

Roaming

The ability to access wireless services outside of one's home calling area. Access can be through one's current provider that offers service in that area or a local provider that has a relationship with one's present service provider.

Rural Service Areas (RSAs)

The FCC divided the country into 428 RSAs and 306 MSAs. Each MSA and RSA was to have two cellular service providers, in order to foster competition. One 25-MHz license was given to the local wireline company (B Block license), and the second was assigned to a nontelephone company (A Block license). The local wireline company was generally the RBOC such as Bell Atlantic or BellSouth.

Smart Message Service (SMS)

The ability to send and receive text messages from a wireless telephone. It was first used in 1992, and is part of the GSM standard. SMS can send/receive up to 160 characters (70 characters non-Latin alphabet) from an SMS-enabled wireless handset.

Smartphones

Handsets that have more on-board processing power than a normal handset. They may also have PIM functionality. This can be enabled by having the Palm OS preloaded on the phone, like the Qualcomm pdQ. These phones are the first foray into a convergence product marrying the benefits of the PDA with the mobile wireless telephone.

Subscriber Information Module (SIM) cards

Removable cards that enable mobile users to customize their handsets and access the services of carriers outside their home region. By inserting an SIM card into an appropriate mobile phone for a region, international travelers can access services from other operators. SIM cards can also contain personal information such as address books, phone numbers, and calendars. They can store account or credit card numbers securely and transfer them from phone to phone, thus enabling m-commerce.

Spectrum

Refers to the ability to transmit signals at a specific frequency and the bandwidth of that frequency.

Switching system

Performs the switching, routing, and call control functions. It also handles billing, call services, and intelligent networking. The key components of the switching system are the Mobile Switching Center, the Home Location Register, the Visitor Location Register, the Authentication Center, and the Equipment Identity Register.

Synchronization

The ability for several devices to work off the same master list, for appointments, contacts, and e-mail without individually having to be updated constantly. To realize the maximum value from synchronization, updates should take place automatically and in real time. Synchronization can be local using wired or wireless technologies. Wired technologies include cables or docking stations, such as for Palm devices, while wireless technologies include Bluetooth, infrared, and 802.11b.

Telematics

Integration of mobile wireless communications and GPS technologies in an automobile. Telematics enables customers to identify their exact location, monitor the status of their vehicles, receive a variety of information and services, and enable theft protection measures. These services can be used by businesses to locate and manage fleets of vehicles, and by individuals for security purposes, emergencies, and access to information available through the call center. All of the major automakers have begun or implemented telematics programs as of 2001.

Telemetry

Refers to machine-to-machine or machine-to-human communications across a wireless network. The main applications of telemetry are monitoring and controlling. Telemetry can be used to monitor the status of vending machines and office equipment or to read utility meters remotely. Telemetry can also be used to control household appliances, security systems, or pipeline switches.

Another promising application for telemetry is credit card authorization. Not only can this be done in a mobile environment, but it is faster than a wired connection, since a dial-up connection does not have to be established in a packet-switched environment.

Terminals

Devices that send and receive radio signals between the subscriber and the wireless network. The primary classes of mobile wireless communication terminals are wireless handsets or telephones, pagers, and PDAs.

Time Distance of Arrival (TDOA)

Uses at least three base stations to measure and compare the arrival time of the control signal from a mobile handset in order to calculate location. To accurately determine location, strict synchronization of the base stations is required. Synchronization is such a critical issue, that an atomic clock is used in each base station. This solution may be attractive to CDMA networks, which are already synchronized, versus GSM networks, which may be asynchronous. TDOA requires line of sight to determine location. This can present problems in rural areas, where three cell sites cannot be accessed simultaneously, and in urban canyons, where multipath reflection can be a problem. TDOA is also less accurate than E-OTD, Cell-ID, or GPS, and can take up to 10 seconds to determine location. TDOA antennas are less expensive and easier to deploy than AOA antennas, and TDOA does not require any handset modifications. However, TDOA is still regarded as a fairly expensive location solution.

Time Division Multiple Access (TDMA)

Utilizes a scheme in which the transmission channel is broken into six time slots. Three of the time slots are used to carry information and three of the time slots are unused in order to minimize interference or noise. Work is currently being done to increase the number of time slots and thus capacity. TDMA increases capacity 3 to 5 times that of analog cellular.

Universal Mobile Telecommunications System (UMTS)

A 3G technology standard (broadband, packet-switched) being developed as a global interoperable system for mobile wireless communications (similar to GSM in Europe). As of 2000, 200 wireless industry members had formed the UMTS Forum to discuss key issues and standards regarding UMTS. UMTS will continue to increase speeds of data transfer (up to 2 Mbps) and allow for additional value-added services. Note that the terms 3G, UMTS, and IMT-2000 are often used interchangeably. In industry parlance, 3G generally refers to packet-switched transmission rates up to 144 kbps in a mobile vehicular environment, 384 kbps in a mobile pedestrian environment, and up to 2 Mbps in a stationary environment. 3G should also provide multimedia capabilities, wireline voice quality and security, and location information, support simultaneous voice and data connections.

Value-added service providers

Firms that seek to use somebody else's network to provide their own service. Examples include WISPs, WASPs, telematics, and telemetry.

Visitor Location Register (VLR)

Database that keeps a record of all mobile subscribers currently active in a particular MSC. The handset routinely sends signals to the VLR to alert the system to its presence. The VLR forwards this information back to the HLR so that calls can be properly routed to the handset.

Web clipping

Proprietary solution developed by Palm to display web content on its PDA. Under this scheme, a template of the website is stored on the Palm device and only the requested information is updated. This procedure drastically cuts down on the amount of data transferred. Full Internet access is not available through Palm.net, but more than 400 websites are currently supported.

Wideband Code Division Multiple Access (W-CDMA)

GSM operators support this as their 3G technology, and it is viewed as the best evolutionary path for GSM/TDMA operators. W-CDMA will support speeds of 384 kbps to 2 Mbps and will utilize a faster chip than CDMA2000. W-CDMA will not be as compatible with early versions of CDMA as CDMA2000, since the W-CDMA chip is set up for the timing of GSM rather than the timing of CDMA. W-CDMA uses a wide channel, 5 MHz, and requires a complete network overhaul. Estimated installation cost for W-CDMA is 100–120% of the original network cost.

Wireless Application Protocol (WAP)

A global standard designed to make Internet services available to mobile users by converting Internet content to a format suitable for the limitations of mobile handsets such as limited screen size and colors, lack of a keyboard or mouse, and limited memory and power. WAP is an enabling wireless Internet technology utilizing wireless markup language (WML) and is more efficient at delivering information than HTML and similar languages.

Wireless Application Service Providers (WASPs)

Companies that host or provision applications in a wireless environment. This is a relatively new industry with new firms entering at a rapid pace. As a new industry, the boundaries are not well defined, and clear winners have not yet emerged. The services offered by firms in this market space include wireless web hosting, wireless web translation, information provision, mobile middleware, and enterprise-application hosting. These services are generally managed from a central location or network operation center.

Wireless handset

Commonly referred to as a cell phone—a device used to receive and transmit voice and data information in a mobile wireless environment.

Wireless Internet Service Providers (WISPs)

Companies that provide Internet access to subscribers and use another company's wireless network to carry their service. This is similar to the wired ISPs, like Earthlink and AOL. Most WISPs also supply a unique portal service that enables customers to personalize their homepage preferences, like a MyYahoo page, and provide a starting point for access to the Internet and a wide range of data.

References

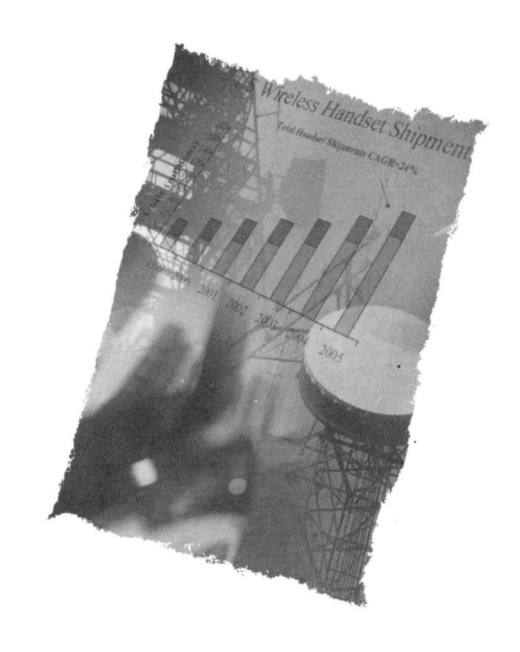

Bensche, John et al. 2000a. *Spectrum auction handbook; An investor's guide to understanding U.S. wireless licensing: Past, present, and future.* Lehman Brothers Research (December 18).

———. 2000b. *OmniSky Corp.* Lehman Brothers Research (November 29).

———. 2000c. *Investing in the wireless location services market.* Lehman Brothers Research (September).

Bernstein Research Call. 2000. *Wireless semiconductors: Battle of the billion dollar press releases.* Bernstein Research Call (February 10).

Bogaty, Lissa and Jeffrey Jamin. 2000. *Telecom equipment in the 21st century.* Donaldson, Lufkin, and Jenrette (September 7).

Ching, Michael and Sandra Zander. 1998. *Wireless scorecard: Update on contract activity.* Merrill Lynch (October 18).

Cowen, S. G. 2000. *Wireless technology: The bumpy but profitable road to 3G* (December).

Crawford, William. 2000. *Handspring Inc.: Hands down an excellent quarter.* Merrill Lynch (October 18).

Derrien, Roger. *Migration paths to 3G.* Lucent Technologies. www.cdg.org/events/exec.

FCC. 2000. *Auctions of licenses in the 747-762 and 777-792 MHz bands scheduled for September 6, 2000, procedures implementing package bidding for auction No. 31.* FCC, Public Notice DA 00-1486 (July 3).

Fleming, Collette et al. 2000. *The urge to merge—2000.* Morgan Stanley Dean Witter (May 22).

Friedland, Peter. 2000. *eWireless enabling the wireless internet.* WR Hambrecht&Co.

Gardner, Richard. 2000. *Palm, inc.: Reading device wars in your future....* Salomon Smith Barney (October 18).

Gillott, Iain. 1999. *Wireless access to the internet, 1999: Everybody's Doin' It.* IDC.

Greiper, S. L. and C. D. Ellingswoth. 2000. *The wireless internet and mobile e-commerce: The second internet revolution.* Commonwealth Associates (October).

House, Jill. 2000. *Market mayhem: The smart handheld devices market forecast and analysis, 1999–2004.* IDC.

IDC. 2000. Burgeoning Bluetooth. *IDC Bulletin.*

Jupiter. 2000. *Mobile revenue models short-term constraints breed long-term opportunities.* Jupiter (April 24).

Labe, Peter. 2000. *Research in motion: Initiating coverage of a growth vehicle.* The Buckingham Research Group (October 12).

Larsen, Christopher et al. 2000. *Wireless communications: Industry update.* Prudential Securities (November 20).

Lundberg, Gregory and Adam Zucker. 2000a. *GoAmerica, inc.: Mobile internet enabler.* Morgan Stanley Dean Witter (November 7).

———. 2000b. *The mobile internet report.* Morgan Stanley Dean Witter (October).

Merrill Lynch. 2000. *Wireless internet more than voice: The opportunity and the issues.* Merrill Lynch (June 5).

Motz, Cynthia et al. 2000. *The wireless edge.* Salomon Smith Barney (October 4).

———. 2000. *The wireless review.* Credit Suisse First Boston.

Munson, Gillian et al. 2000. *Palm tops estimates despite component shortages.* Morgan Stanley Dean Witter (September 27).

Pottorf, Callie. 2000. *Pulling location into wireless: Location-based service market forecast and analysis, 1999–2004.* IDC (May).

———. 1999a. Cellular/PCS distribution channels update, 1998–2003. *IDC Bulletin* (November).

———. 1999b. Everybody churn, churn, churn: Cellular/PCS churn, 1998–2003. *IDC Bulletin* (September).

———. 1999c. Prepaid cellular/PCS: Snowballing success? *IDC Bulletin.*

Pottorf, Callie and Charul Vyas. 2000. *U.S. wireless services and devices market assessment, 1999–2004.* IDC (May).

Redman, P. 2000. *Paging operators fights to remain relevant.* Gartner Group (October 19).

Rietman, Julie. 1998. *Worldwide cellular and PCS infrastructure market assessment, 1997–2002.* IDC (November).

Roberts, Mark A. and Michael A. Whitfield. 2000. *Look ma, no wires.* First Union Securities (June 3).

Robinson-Humphrey. 2000. *The mobile internet.* Robinson-Humphrey.

Rollins, Michael I. 2000. *The wireless edge.* Salomon Smith Barney (October 4).

Strategy Analytics. 2000a. *Cellular service trends industry report.* Strategy Analytics (June). www.strategyanalytics.com.

——— 2000b. *Global cellular data applications, revenue & 3G migration.* Strategy Analytics (December).

———. 2000c. *U.S. cellular handset market status.* Strategy Analytics (October).

———. 2000d. *U.S. wireless voice market forecast (2000–2005).* Strategy Analytics (February).

———.2000e. *Wireless telemetry: Real market opportunity or niche application?* Strategy Analytics (March).

———. 1999. *U.S. paging industry: Market update and forecast.* Strategy Analytics (October).

———. 1998. *U.S. wide area paging market forecast (1998–2003).* Strategy Analytics (March).

Telematics growth uneven, winners uncertain. *Global Positioning & Navigation News* (November 29, 2000).

Vyas, Charul. 2000. *Personal wireless communications user survey, 2000.* IDC (April).

———. 2000b. *Wireless ASPs—who is, who isn't, who wants to be.* IDC (October).

Index

A

D

E

F

G

H

I

K

L

M

N

O

P

Q

R

S

T

U

V

W

About the Author

John P. Burnham has extensive experience in the telecommunications field. He is an executive with responsibility for North American strategy development, competitive analysis, and research for one of the world's preeminent consumer electronics companies. Prior to that, he was Manager of Corporate Strategy for a Regional Bell Operating Company.

Before working in telecommunications, Mr. Burnham spent several years with one of the world's largest automobile manufacturers, in a number of roles of increasing responsibility, including Pricing, Incentives, Mergers and Acquisitions, and Strategic and Business Planning.

Mr. Burnham has an MBA from the Fuqua School of Business at Duke University, with a concentration in Strategic Planning and Marketing. He also earned a B.S. in Finance from Arizona State University.

Mr. Burnham can be contacted at jpburnham@hotmail.com.